Lecture Notes in Computer Science 5546

Commenced Publication in 1973
Founding and Former Series Editors:
Gerhard Goos, Juris Hartmanis, and Jan van Leeuwen

W0080121

Lecture Notes in Computer Science 5510

Hans van den Berg   Geert Heijenk
Evgeny Osipov   Dirk Staehle (Eds.)

Wired/Wireless Internet Communication

7th International Conference, WWIC 2009
Enschede, The Netherlands, May 27-29, 2009
Proceedings

 Springer

Volume Editors

Hans van den Berg
TNO Information and Communication Technology
2600 GB Delft, The Netherlands
E-mail: J.L.vandenBerg@tno.nl

Geert Heijenk
University of Twente, Faculty of Electrical Engineering,
Mathematics and Computer Science
7500 AE Enschede, The Netherlands
E-mail: geert.heijenk@utwente.nl

Evgeny Osipov
Luleå University of Technology, Department of Computer Science
and Electrical Engineering
97187 Luleå, Sweden
E-mail: Evgeny.Osipov@ltu.se

Dirk Staehle
University of Würzburg, Department of Distributed Systems
D-97074 Würzburg, Germany
E-mail: dstaehle@informatik.uni-wuerzburg.de

Library of Congress Control Number: Applied for

CR Subject Classification (1998): C.2, D.2, D.4.4, H.4, H.3.5, K.6.4

LNCS Sublibrary: SL 5 – Computer Communication Networks
and Telecommunications

ISSN   0302-9743
ISBN-10   3-642-02117-4 Springer Berlin Heidelberg New York
ISBN-13   978-3-642-02117-6 Springer Berlin Heidelberg New York

springer.com

© Springer-Verlag Berlin Heidelberg 2009
Printed in Germany

Typesetting: Camera-ready by author, data conversion by Scientific Publishing Services, Chennai, India
Printed on acid-free paper   SPIN: 12689521   06/3180   5 4 3 2 1 0

Preface

The seventh edition of the International Conference on Wired/Wireless Internet Communications (WWIC) was organized by the University of Twente in May 2009. Since the first event in 2002, WWIC has been established as a highly selective conference focussing on the rapidly developing field of wireless networking, and providing an international forum for the presentation and discussion of cutting-edge research in the field.

The WWIC 2009 call for papers attracted 39 submissions from 20 countries, which were subject to thorough review by the Technical Program Committee members and additional experts. The selection process resulted in the acceptance of 13 papers, organized into 4 technical sessions. The major themes of WWIC this year were energy efficiency, security, reliability, and routing protocols in wireless sensor and ad hoc networks as well as handover and mobility management in heterogeneous environments. We are grateful to Matthias Grossglauser (Nokia Research Center, Finland, and EPFL Lausanne, Switzerland), who accepted our invitation to give the WWIC 2009 keynote speech. Further, we would like to thank Hans Appel (Sun Microsystems, The Netherlands), and Remco Litjens (TNO ICT, The Netherlands) for giving invited presentations at this year's event. In addition to the main technical program, the third ERCIM workshop on eMobility took place on the first day of WWIC 2009.

We thank the authors for choosing WWIC 2009 as the conference to submit their results to. We would also like to thank all the members of the Technical Program Committee, as well as the additional reviewers for their effort in providing detailed and constructive reviews. The support of Springer LNCS is gratefully acknowledged again this year. We would also like to thank our sponsors, in particular, The Netherlands Organisation for Scientific Research (NWO), the Centre for Telematics and Information Technology (CTIT) of the University of Twente, the European Research Consortium for Informatics and Mathematics (ERCIM), Nokia, and TNO. We are grateful to the members of the Local Organizing Committee for their efforts.

We hope that all attendees enjoyed the scientific and social program as well as the beautiful campus of the University of Twente and the surrounding region. We look forward to welcoming you at WWIC 2010, which will be held in Luleå, Sweden!

May 2009

Hans van den Berg
Geert Heijenk
Evgeny Osipov
Dirk Staehle

Organization

Executive Committee

General Chairs   Geert Heijenk, University of Twente,
         The Netherlands
         Evgeny Osipov, Luleå University of Technology,
         Sweden
TPC Chairs   Hans van den Berg, TNO ICT, The Netherlands
         Dirk Staehle, University of Würzburg, Germany

Local Organizing Committee

Desislava Dimitrova   University of Twente, The Netherlands
Silvia Meijran   University of Twente, The Netherlands

Steering Committee

Torsten Braun   University of Bern, Switzerland
Georg Carle   Universität München, Germany
Giovanni Giambene   University of Siena, Italy
Yevgeni Koucheryavy Tampere University of Technology, Finland
Peter Langendoerfer   IHP Microelectronics, Germany
Ibrahim Matta   Boston University, USA
Vassilis Tsaoussidis   Demokritos University, Greece
Nitin Vaidya   University of Illinois, USA

Supporting and Sponsoring Organizations

Centre for Telematics and Information Technology
ERCIM
Nokia
Netherlands Organisation for Scientific Research (NWO)
TNO
University of Twente

Technical Program Committee

Onur Altintas   Toyota InfoTechnology Center, Japan
Ozgur B. Akan   Middle East Technical University, Turkey
Manuel
  Alvarez-Campana Universidad Politecnica de Madrid, Spain

Leonardo Badia   IMT Lucca, Italy
Sergey Balandin   Nokia, Finland
Mortaza Bargh   Telematics Institute, The Netherlands
Carlos Bernardos   Universidad Carlos III de Madrid, Spain
Bharat Bhargava   Purdue University, USA
Fernando Boavida   University of Coimbra, Portugal
Thomas Michael Bohnert   SAP Research, Switzerland
Sem Borst   University of Twente, The Netherlands
Richard Boucherie   University of Twente, The Netherlands
Torsten Braun   University of Bern, Switzerland
Rafaelle Bruno   IIT-CNR, Italy
Wojciech Burakowski   Warsaw University of Technology, Poland
Maria Calderon   Universidad Carlos III de Madrid, Spain
Xiuzhen Cheng   George Washington University, USA
Bong Dae Choi   Korea University, Korea
Mieso Denko   University of Guelph, Canada
Michel Diaz   LAAS-CNRS, France
Magda El Zarki   University of California, Irvine, USA
Giovanni Giambene   University of Siena, Italy
Jarmo Harju   Tampere University of Technology, Finland
Geert Heijenk   University of Twente, The Netherlands
Markus Hofmann   Bell Labs / Alcatel-Lucent, USA
Haruki Izumikawa   KDDI R&D Laboratories, Japan
Yuming Jiang   Norwegian University of Science and Technology,
                              Norway
Andreas Kassler   Karlstad University, Sweden
Ibrahim Khalil   RMIT University, Australia
Byung Kim   University of Mass. Lowell, USA
Yevgeni Koucheryavy   Tampere University of Technology, Finland
Rolf Kraemer   IHP Microelectronics, Germany
Peter Kropf   University of Neuchâtel, Switzerland
Dirk Kutscher   University of Bremen, Germany
Peter Langendoerfer   IHP Microelectronics, Germany
Kenji Leibnitz   Osaka University, Japan
Leszek Lilien   Western Michigan University, USA
Remco Litjens   TNO ICT, The Netherlands
Hai Liu   Hong Kong Baptist University, Hong Kong
Pascal Lorenz   University of Haute Alsace, France
Andreas Mäder   NEC Labs, Germany
Christian Maihöfer   Daimler AG, Germany
Lefteris Mamatas   Demokritos University, Greece
Saverio Mascolo   Politecnico di Bari, Italy
Enzo Mingozzi   University of Pisa, Italy
Dmitri Moltchanov   Tampere University of Technology, Finland
Edmundo Monteiro   University of Coimbra, Portugal
Liam Murphy   University College Dublin, Ireland

Marc Necker	Universität Stuttgart, Germany
Qiang Ni	Brunel University, UK
Ioanis Nikolaidis	University of Alberta, Canada
Guevara Noubir	Northeastern University, USA
Evgeny Osipov	Luleå University of Technology, Sweden
Philippe Owezarski	LAAS-CNRS, France
George Pavlou	University College London, UK
Utz Roedig	Lancaster University, UK
Theodoros Salonidis	Thomson - Paris Research Labs, France
Guenter Schaefer	TU Ilmenau, Germany
Jochen Schiller	Free University Berlin, Germany
Patrick Sénac	ISAE, France
Dimitrios Serpanos	University of Patras, Greece
Vasilios Siris	University of Crete and ICS-FORTH, Greece
Dirk Staehle	University of Würzburg, Germany
Burkhard Stiller	University of Zurich and ETH Zurich, Switzerland
Phuoc Tran-Gia	University of Würzburg, Germany
Vassilis Tsaoussidis	Demokritos University, Greece
Hans van den Berg	TNO ICT / University of Twente, The Netherlands
Rob van der Mei	Centre for Mathematics and Computer Science, The Netherlands
Piet Van Mieghem	Technical University of Delft, The Netherlands
Alexey Vinel	Saint Petersburg Institute for Informatics and Automation, Russia
Miki Yamamoto	Kansai University, Japan
Evsen Yanmaz	University of Klagenfurt, Austria
Chi Zhang	Juniper Networks, USA
Martina Zitterbart	University of Karlsruhe, Germany

Additional Reviewers

Anwander, Markus	Kuipers, Fernando	Psaras, Ioannis
Beben, Andrzej	Latré, Benoît	Skordoulis, Dionysios
Doerr, Christian	Mitoraj, Piotr	Staehle, Barbara
Hurni, Philipp	Mueller, Christian	Wagenknecht, Gerald
Koutsogiannis, Efthymios	Papastergiou, Giorgos	Zarakovitis, Charilaos
	Pries, Rastin	

Table of Contents

Energy Efficient WSN Design

A Novel MAC Protocol for Event-Based Wireless Sensor Networks:
Improving the Collective QoS...................................... 1
*Cristina Cano, Boris Bellalta, Jaume Barceló, and
Anna Sfairopoulou*

Implicit Sleep Mode Determination in Power Management of
Event-Driven Deeply Embedded Systems........................... 13
André Sieber, Karsten Walther, Stefan Nürnberger, and Jörg Nolte

On Prolonging Sensornode Gateway Lifetime by Adapting Its Duty
Cycle.. 24
Marcin Brzozowski and Peter Langendoerfer

Routing and Transport Protocols for WSNs

Routing and Aggregation Strategies for Contour Map Applications in
Sensor Networks 36
Shoudong Zou, Ioanis Nikolaidis, and Janelle Harms

Path-Based Reputation System for MANET Routing 48
Ji Li, Teng-Sheng Moh, and Melody Moh

Hop-to-Hop Reliability in IP-Based Wireless Sensor Networks - A
Cross-Layer Approach 61
Gerald Wagenknecht, Markus Anwander, and Torsten Braun

Security and Protocol Design

A Scalable Security Framework for Reliable AmI Applications Based
on Untrusted Sensors 73
*José M. Moya, Juan Carlos Vallejo, Pedro Malagón, Álvaro Araujo,
Juan-Mariano de Goyeneche, and Octavio Nieto-Taladriz*

The Quest for Mobility Models to Analyse Security in Mobile Ad Hoc
Networks 85
*Mauro Conti, Roberto Di Pietro, Andrea Gabrielli,
Luigi Vincenzo Mancini, and Alessandro Mei*

RDTN: An Agile DTN Research Platform and Bundle Protocol
Agent 97
Janico Greifenberg and Dirk Kutscher

Mobility and Handover Management

Realization Aspects of Multi-Radio Management Based on
IEEE 802.21 109
   Christian M. Mueller, Harald Eckhardt, and Rolf Sigle

Seamless Mobility of Senders Transmitting Multi-user Sessions over
Heterogeneous Networks 121
   Luis Veloso, Eduardo Cerqueira, Paulo Mendes, and
   Edmundo Monteiro

An Adaptive Optimized RTO Algorithm for Multi-homed Wireless
Environments 133
   Sheila Fallon, Paul Jacob, Yuansong Qiao, and Liam Murphy

Handover Incentives for WLANs with Overlapping Coverage 146
   Xenofon Fafoutis and Vasilios A. Siris

Author Index 159

A Novel MAC Protocol for Event-Based Wireless Sensor Networks: Improving the Collective QoS

Cristina Cano, Boris Bellalta, Jaume Barceló, and Anna Sfairopoulou

Departament de Tecnologies de la Informació i les Comunicacions, (DTIC)
Universitat Pompeu Fabra,
Passeig de la Circumval.lació, 8 08003 Barcelona, Spain
{cristina.cano,boris.bellalta,
jaume.barcelo,anna.sfairopoulou}@upf.edu
http://www.nets.upf.edu/

Abstract. WSNs usually combine periodic readings with messages generated by unexpected events. When an event is detected by a group of sensors, several notification messages are sent simultaneously to the sink, resulting in sporadic increases of the network load. Additionally, these messages sometimes require a lower latency and higher reliability as they can be associated to emergency situations. Current MAC protocols for WSNs are not able to react rapidly to these sporadic changes on the traffic load, mainly due to the duty cycle operation, adopted to save energy in the sensor nodes, resulting in message losses or high delays that compromise the event detection at sink. In this work, two main contributions are provided: first, the collective QoS definitions are applied to measure event detection capabilities and second, a novel traffic-aware Low Power Listening MAC to improve the network response to sporadic changes in the traffic load is presented. Results show that the collective QoS in terms of collective throughput, latency and reliability are improved maintaining a low energy consumption at each individual sensor node.

Keywords: WSNs, Collective QoS, MAC, B-MAC.

1 Introduction

In a WSN (Wireless Sensor Network) a group of nodes communicate environmental data to a central device, called sink [1]. Sensor nodes are formed by the sensor unit (that detects the data) and the communication module (that sends the data wirelessly). Usually, WSNs deployments are large and not easily accessible (for instance, a forest deployment). These two main characteristics impose several constraints on the sensor nodes: devices need to be small and inexpensive implying limited power sources, memory and processing resources. Thus, the most critical concern is the energy consumption, which must be minimized to achieve the expected network lifetime. Among all the tasks of a sensor

H. van den Berg et al. (Eds.): WWIC 2009, LNCS 5546, pp. 1–12, 2009.
© Springer-Verlag Berlin Heidelberg 2009

node, communicating the data is the most consuming one due to the transceiver power consumption. Apart from the transceiver design, the Medium Access Control (MAC) protocol is the key factor influencing the transceiver operation [2]. This is the reason of the large amount of MAC protocols specially designed for WSNs. The common approach to reduce the energy consumption is to periodically put the transceiver into sleep mode, working in a low duty cycle operation, listening to the channel only a small percentage of time. Thus, this solution is able to increase the battery duration of the sensor nodes but degrades the network performance.

Usually, WSNs applications do not require QoS (Quality of Service) in terms of strict guarantees for the network delay and/or packet losses, specially in those WSNs dedicated to a simple monitoring task of a stable system. However, in an event-based WSN the messages related to the events occurrence require QoS guarantees in order to increase the probability that the event is properly detected at sink. Notice that, in this case, the QoS is related to the events (event detection delay, event detection reliability, etc.) and not to individual messages. To this aim, collective QoS is defined as the QoS in terms of throughput, latency, reliability and packet loss of the set of messages related to a certain event. Event-based common applications include target tracking, emergency detection and disaster relief among others.

There are several proposed MAC protocols which try to provide traditional QoS as well as low energy consumption. However, to the best of our knowledge, there are not any MAC protocol designed to provide collective QoS in WSNs as a primary objective. We will show that the protocol presented here improves the defined collective QoS metrics maintaining the energy consumption as a strong constraint, as it is designed to only react to the sporadic increases of traffic load due to the messages caused by an event detection.

The rest of the paper is organized as follows: the definitions of the collective QoS metrics are provided in Section 2, while in Section 3 an overview of the different MAC protocols for WSNs is presented, including those that take QoS into account. In Section 4, the protocol to increase the collective QoS in event-based WSNs is described and its results are discussed in Section 5. Finally, some concluding remarks are given.

2   QoS in WSNs: A Collective Approach

Due to the high energy constraint in WSNs, QoS (understood as in traditional communication networks) has not been given enough attention, as in most cases, the techniques used to reduce the energy consumption conflict with those used to guarantee the traditional QoS, such as the duty cycle operation at the MAC layer. Moreover, like other issues in WSN, QoS needs are highly application dependent. For instance, periodic readings usually do not need a strict grade of QoS since this kind of applications are normally non sensitive to high delays and can tolerate a certain percentage of packet loss. However, event-based applications need that the event-related messages arrive at sink with a certain grade of QoS. Notice that the QoS requirements differ from the traditional end-to-end ones where the metrics are measured packet by packet. The QoS should now be measured in a data-centric way, which is known as collective QoS. Collective QoS is defined as the QoS (delay, bandwidth, packet loss, etc.) of the set of packets related to a specific event [3]; i.e. it is not important the delay of the individual messages but it is crucial the latency from the event generation until the event detection. Here, we extend the collective QoS metrics to: collective delay, bandwidth, packet loss and reliability.

Collective Delay (CD) refers to the time span between the event occurrence and the event detection at sink, assuming that the sink needs to receive N packets referring to an event to ensure the event occurrence. In this work, it is considered that the sink detects the event when it receives N packets related to it, independently of the time span. Therefore, the CD of event i is defined as follows [4]:

$$CD_i = R_i(N) - \min\left(D_i(j)\right), \; j \; \epsilon \; \mathbf{E}_i \tag{1}$$

where $R_i(N)$ is the instant at which the Nth message of event i reaches the sink, $(D_i(j))$ is the moment at which the j sensor detects event i and $\mathbf{E}_i$ is the group of sensors that detect the ith event.

Similarly, the Collective Bandwidth (CB) can be defined as the bandwidth required to detect an event:

$$CB_i = \frac{N \cdot l_{data}}{CD_i} \tag{2}$$

where l_{data} is the packet length.

The number of packets lost during the information delivery period, Collective Packet Loss (CPL), can be defined as follows:

$$CPL_i = L_i(T), \; T = [\min\left(D_i(j)\right), \; R_i(N)], \; j \; \epsilon \; \mathbf{E}_i \tag{3}$$

where L_i is the number of packets lost of event i in the time interval T (from event detection by the sensor node to event detection at sink).

The number of packets lost can be linked to the Collective Reliability (CR), which is the fraction of correctly detected events among all events generated (G).

$$CR = \frac{1}{G} \cdot \sum_{k=1}^{G} I(k), \; I(k) = \begin{cases} 1 & \text{if } k : (|\mathbf{E}_i| - TPL_i \geq N) \\ 0 & \text{otherwise} \end{cases} \tag{4}$$

where TPL_i is the total number of packets lost related to event i and $|.|$ denotes the number of elements (cardinality) of the set.

3 MAC Protocols for WSNs

In the last decades several advances related to Wireless Networks have been made. In this sense, it is important to point out the great success of the IEEE 802.11 [5] standard that defines the physical and medium access control layers of a wireless node. The channel sharing method is basically a CSMA/CA technique in which nodes sense the channel and wait a random time before transmitting, reducing the number of collisions. Unfortunately, these protocols are not suitable for WSNs, mainly due to the limited energy resources available in sensor devices. Energy consumption of a WSN occurs in three different domains: sensing, data processing and communicating. Among these, radio communication is the major consumer of energy. In traditional wired and wireless MAC protocols nodes are always listening to the channel in order to receive possible transmissions, but in WSNs this feature will deplete the battery of sensor nodes rapidly. Apart from that, control messages, carrier sensing and acknowledgements, common in traditional MAC protocols, become noticeable overhead if compared to the small data payloads found in most WSNs. To reduce the energy consumption, these sources of energy waste should be reduced [2]:

- Collisions: When a collision occurs none of the packets involved can be correctly received. Therefore, retransmissions that cause an extra energy consumption are needed.
- Idle Listening: Listening to the medium when there is nothing to receive. This effect has been identified as the major energy waste in WSNs [6].
- Overhearing: To receive a message from the wireless channel destined to another recipient.
- Overhead: Control messages and extra information in the data packet.

3.1   Common Energy-Aware MAC Protocols

In order to reduce the energy consumption in WSNs the most common approach is to implement a low duty cycle MAC protocol that combines listen with sleep intervals. These protocols can be classified into two main categories: scheduled protocols (in which some kind of organization between nodes is made in order to decide when to sleep) and unscheduled (where each node independently selects its own schedule)[1].

Scheduled Protocols. Scheduled MAC protocols [2] reduce the energy waste by coordinating the sensor nodes with a common schedule (when to listen and sleep). In this kind of protocols nodes know when their neighbours will be awake to receive messages. Having an organization to access the channel limits the idle listening periods and the overhearing. The drawback of these mechanisms is the cost to create and maintain the schedule. Apart from that, synchronization becomes a problem because periodic beacons should be used or a higher precision oscillator has to be included in the sensor thereby increasing its cost.

The most representative and probably the most studied MAC protocol for WSNs is S-MAC [6]. In S-MAC, neighbouring nodes are synchronized together to the same schedule in order to reduce control overhead. The network is then divided into virtual clusters of nodes synchronized together. Nodes broadcast

---

[1] Here the terms *scheduled* and *unscheduled* refer to the organization of the listen and sleep intervals not in the access to the channel.

synchronism packets to allow neighbours to learn their schedules. Therefore, all nodes can communicate with all their neighbours although they have different schedules. This technique allows to reduce the energy waste due to idle listening. To avoid collisions, nodes use physical and virtual carrier sense and the RTS/CTS mechanism.

In this category it can also be classified the IEEE 802.15.4 [7], a standard for small devices with low power resources, low data rates requirements and basically designed for nearby communications. It defines two different types of access methods: the beacon-enabled and the nonbeacon-enabled modes. In the beacon-enabled mode the channel time is divided into an active part (formed by reserved slots and a contention access) and an inactive part (where the network coordinator can go to sleep). In the nonbeacon-enabled mode the channel access is based on an unslotted CSMA/CA where the coordinator is unable to sleep. Both modes allow end devices to sleep to save energy and wake up periodically to send and/or to poll the coordinator for data. The standard defines the star and peer-to-peer topologies, although it is focused in the star topology leaving some features of the peer-to-peer topology undefined.

Unscheduled Protocols. On the other hand, unscheduled or random MAC protocols [2] have the advantage of their simplicity. Since there are no schedules to be maintained or shared, the node consumes fewer processing resources, it requires a smaller memory and the number of messages that have to be transmitted is reduced. Moreover, they are more flexible to support different types of traffic loads. However, there exist idle listening and overhearing.

The most representative protocol is B-MAC [8], in which each node selects a sleep schedule independently of the neighbourhood. Each time a node wants to send a packet, it first sends a preamble long enough to be listened by the intended recipient. To guarantee that the preamble and the listening period of the recipient overlap, the length of the preamble has to be equal to the duty cycle. If a node detects activity on the channel while it is listening, it will remain awake to receive the packet. Messages can be immediately transmitted if the carrier sense determines that the medium is idle. However, the long preamble transmission increases the energy spent in overhead, overhearing and the latency to send a packet.

Hybrid Protocols. As far as the authors know there is not any hybrid protocol that combines scheduled and unscheduled accesses. In this work a traffic-aware self-adaptive MAC protocol that combines the benefits of both approaches is presented. In low load conditions the unscheduled access is used while, as the load increases, a scheduled access after transmissions is adopted without requiring additional signalling for schedule creation and maintenance.

3.2 QoS-Aware MAC Protocols for WSNs

There are not any MAC protocol that considers collective QoS metrics. Those which are focused on QoS are based on end-to-end traditional QoS metrics instead. For instance, PQ-MAC [9] is a slotted protocol that defines different types of priority to access the channel. It consumes less energy than S-MAC but it suffers from a costly setup phase to assign a slot to each node. Q-MAC [10] defines multiple queues on each sensor node to provide different service levels and uses different contention windows depending on the packet criticality among other metrics. Finally, RL-MAC [11] adapts the duty cycle of S-MAC based on the inferred state of the other nodes and defines three different service levels with different contention windows.

4   LWT-MAC: LPL with Scheduled Wake Up after Transmissions

To improve the collective QoS, the MAC protocol should react to the sporadic increases in the network load due to an event detection without compromising the energy consumption. The protocol presented here combines an unscheduled access (used under low load) and a scheduled access (used at high load). The unscheduled access is based on the Low Power Listening (LPL) B-MAC [8] protocol and a variation of S-MAC [6] is used in the scheduled access. With low traffic load, B-MAC shows better performance as the energy consumption and delay are reduced. However, as traffic grows, the continuous collision of preambles, specially in a multihop network, significantly degrades its performance. On the other hand, S-MAC with adaptive duty cycle can provide low delay, low energy consumption and a better hidden terminal management in high load conditions.

The LPL with scheduled Wake up after Transmissions (LWT-MAC) extends the normal operation of B-MAC by taking advantage of the local synchronization of all nodes that overhear a transmission. The scheduled wake up after transmissions wakes up overhearing nodes simultaneously at the end of the ongoing transmission in order to send or receive packets without requiring the transmission of the long preamble (see Fig. 1). In this scheduled phase, RTS/CTS messages are mandatory; they are used to allow overhearing nodes to go to sleep until the end of the current transmission by setting the Network Allocation Vector (NAV) timer to its duration. It is assumed that all nodes wake up immediately after a transmission, although a sleep period can be added to save more energy following the S-MAC design [6]. To prevent collisions, nodes wait a

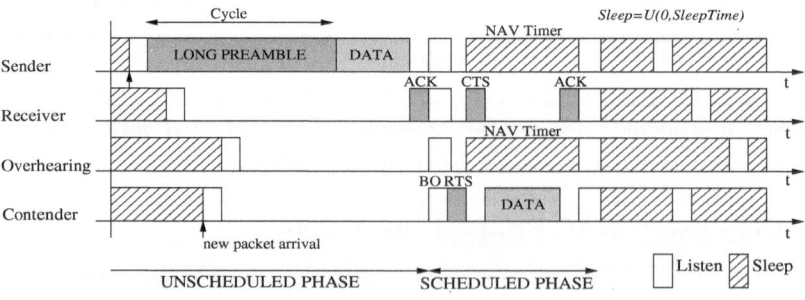

Fig. 1. LPL with scheduled wake up after transmissions

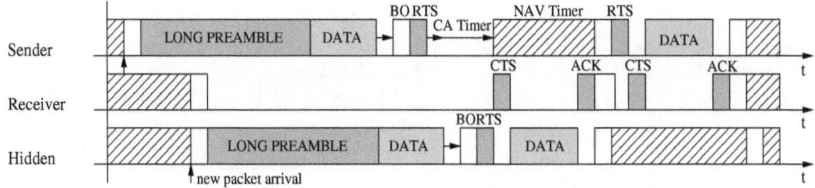

Fig. 2. Retransmission procedure in the unscheduled access

random backoff (BO) before sending an RTS packet, then, the listen time after transmissions should be equal to the maximum backoff. In the case that nodes wake up after a transmission and nothing is received, they go to sleep during a random time (with maximum value equal to the sleep time) moving towards the unscheduled phase.

If the transmission of a packet fails, the protocol begins the retransmission procedure that differs depending on which phase it occurs:

Retransmission procedure in the unscheduled access. If a transmission failure occurs during the unscheduled access (the ACK is not received), the sender retransmits the packet by sending an RTS after waiting a random backoff. As the transmission failure can be caused by collisions of preambles, the retransmission increases the probability to receive the RTS correctly. After that, in case the CTS is not received (there are two consecutive transmission failures), it is assumed that the intended recipient is either involved in another transmission or waiting for a transmission to finish. In order to alleviate consecutive collisions, the sender does not immediately retransmit the message. Instead, it waits a Collision Avoidance (CA) timer (set to the duration of a preamble and a frame transmission). During the CA timer the node keeps listening to the channel, that provides the opportunity to get synchronized with the current ongoing transmission (if any) after overhearing a related message. If that is the case, it can sleep until the transmission finishes and retry transmission using the scheduled method (see Fig. 2). Otherwise, if the CA timer expires, the node retries transmission using the long preamble again.

Retransmission procedure in the scheduled access. If a transmission fails during the scheduled access it can be either because the CTS or the ACK are not received. In case the RTS fails (CTS not received), it is assumed that the recipient has not overheard the past transmission and, therefore, it is sleeping (notice that the long preamble has not been previously sent). In this case, the node will retry to send the message by sending the long preamble first in order to wake up the receiver (see Fig. 3). If the packet transmission is not acknowledged the sender waits a CA timer as previously described in the unscheduled access.

Note that at instantaneous increases of the network load (for instance at events occurrence) the protocol reduces the time to send a packet and provides a better hidden terminal management if compared to B-MAC. This allows to increase the collective QoS of the event-based messages.

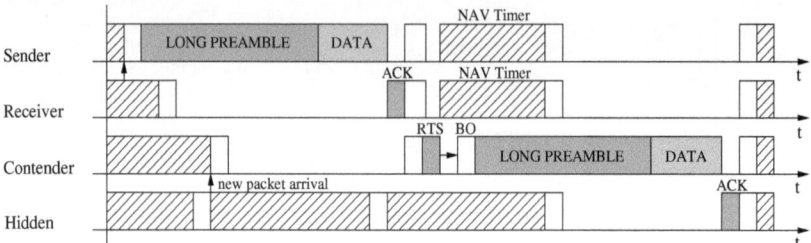

Fig. 3. Retransmission procedure in the scheduled access

5   Results

To evaluate the performance of the presented protocol, the SENSE [12] simulator has been used. The LPL extension of the physical layer has been implemented as well as the compared MAC protocols: IEEE 802.11 (used as a reference only), B-MAC and the proposed LPL with Wake up after Transmissions (LWT-MAC)[2]. Additionally, to simulate a multihop network avoiding the effects of the routing protocol[3] the Floyd algorithm has been used to compute the shortest path between any pair of nodes. Finally, an event generator connected to the application layer of each node has been implemented. It generates random events and notifies those sensor nodes that are inside the coverage radius of the event in order to send event-based messages to the sink.

5.1   Scenario

The considered scenario (see Fig. 4) is a multihop event-based WSN with 100 nodes randomly placed in a $100 \times 100m^2$ area. The radio range of each node is 43m. All simulations have a duration of 500,000s. For a fair comparison, the RTS/CTS mechanism has been used in all MAC protocols. In the case of B-MAC and LWT-MAC, the RTS is immediately sent after the long preamble. Each sensor node generates two kind of traffic profiles: *i)* periodic messages generated following a Poisson distribution and *ii)* event-based messages that are generated if the sensor node is inside the coverage radius of the event. The offered load coming from the periodic data is changed while the number of events generated and their positions are maintained for all the simulations. Event positions are selected randomly inside the area and they have a constant coverage radius of 30m. The time between events follows an exponential distribution with mean 600s and N (the number of messages that are required to detect an event at sink) has been set to 5. Table 1 shows the parameters used in the simulation,

---

[2] Performance results of the original S-MAC have not been included in the comparison as it provides worse results in throughput, energy consumption and delay compared to B-MAC [8].

[3] Static routes are created at the beginning of the simulation avoiding route management traffic to influence the results.

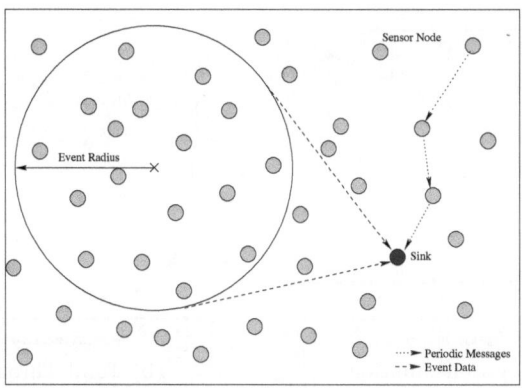

Fig. 4. Considered WSN scenario

Table 1. Simulation Parameters

Parameter	Value	Parameter	Value
Data Rate	20kbps	Listen/Sleep Time	24.5/75.5ms
Slot	1ms	Data Packet Size (l_{data})	30bytes
DIFS	10ms	Control Packet Size	8bytes
SIFS	5ms	Tx Consumption	24.75mW
Retry Limit	5	Rx/Idle Consumption	13.5mW
Queue Length	10pkts	Sleep Consumption	15μW
CWmin (B-MAC)	64	CWmin (802.11)	32
CWmax (B-MAC)	64	CWmax (802.11)	1024

most of them have been extracted from the ns-2 implementation of S-MAC [13] and from the B-MAC specification [8].

5.2 Performance Results

Fig. 5 and Fig. 6 show the performance results with different packet interarrival values for the periodic data. It can be seen that the energy consumption (Fig. 5a) of the LWT-MAC is slightly higher if compared to the B-MAC protocol except for high traffic loads. This is caused by the listen after transmissions mechanism, which at low traffic loads results in unnecessary listening times after transmissions as the probability that a neighbour node (including the node that has transmitted) has a packet ready to be transmitted is low. However, when traffic load increases, this probability is higher, and the listen after transmissions mechanism moves the network to the scheduled access, allowing to send messages without sending the long preamble and reducing the number of collisions. If compared to the IEEE 802.11 the energy reduction is noticeable.

Regarding the traditional metrics, throughput (Fig. 5b), delay (Fig. 5c and Fig. 5e) and reliability (Fig. 5d and Fig. 5f) for periodic and individual event-based messages, the results show that the LWT-MAC protocol provides a

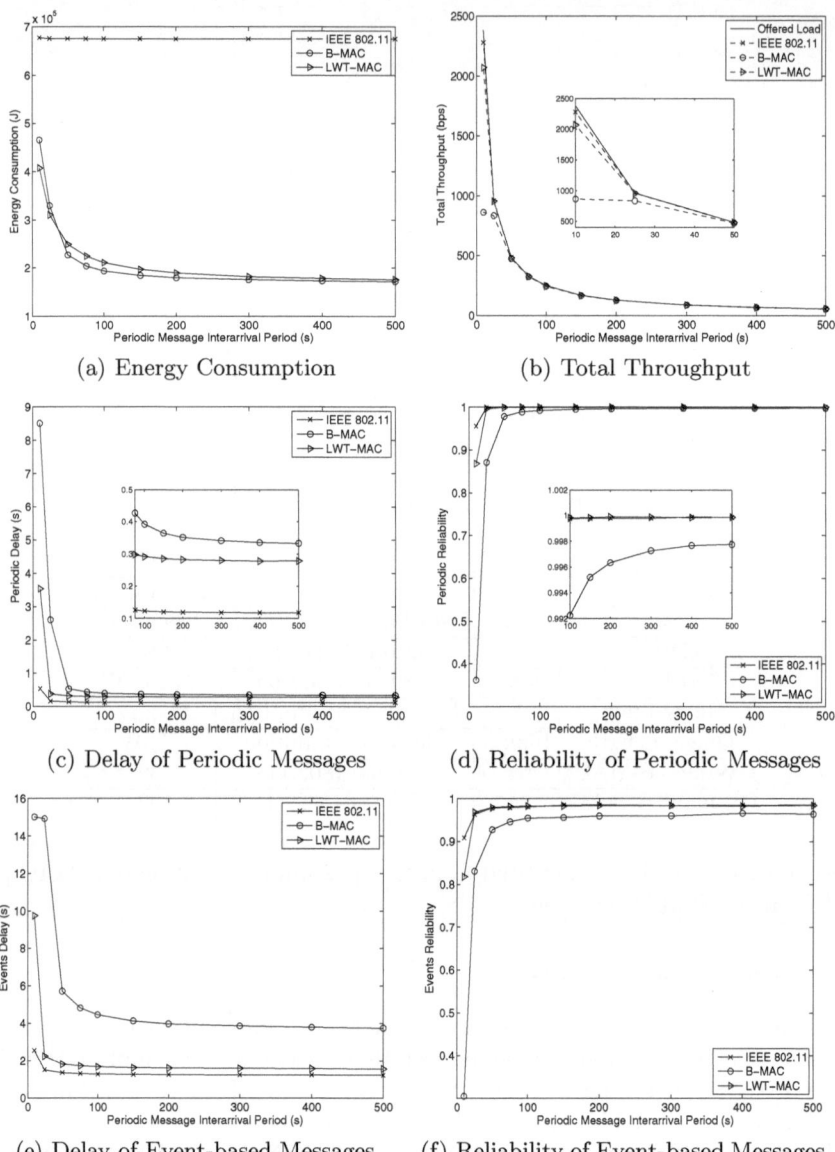

(a) Energy Consumption   (b) Total Throughput

(c) Delay of Periodic Messages (d) Reliability of Periodic Messages

(e) Delay of Event-based Messages (f) Reliability of Event-based Messages

Fig. 5. Performance results of the multihop WSNs scenario with periodic and event-based traffic profiles

performance similar to the IEEE 802.11 protocol, delivering smaller delay (more noticeable in event-based messages) and higher throughput and reliability than B-MAC.

Fig. 6a and Fig. 6b as well as Table 2 show the collective delay, bandwidth and reliability metrics measured from the event-based messages of those events

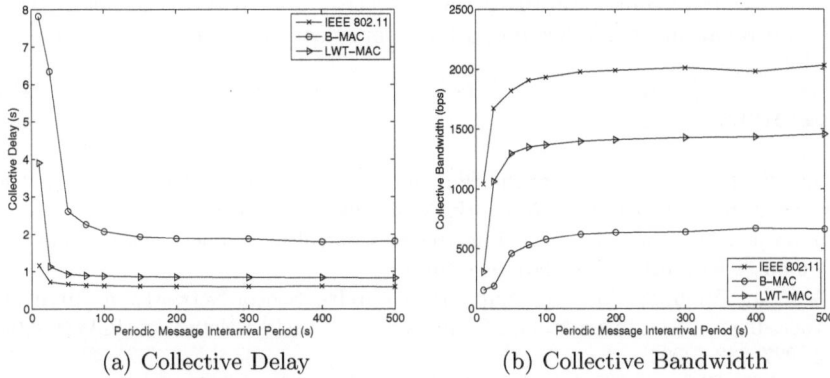

(a) Collective Delay   (b) Collective Bandwidth

Fig. 6. Collective metrics of the multihop WSNs scenario with periodic and event-based traffic profiles

correctly detected. In Fig. 6a it can be seen how the LWT-MAC improves the collective delay if compared to the B-MAC protocol, showing a performance close to the IEEE 802.11. In Fig. 6b, where the collective bandwidth is depicted, LWT-MAC also shows a better performance compared to B-MAC. Regarding the collective reliability (Table 2) a significant improvement is found with the LWT-MAC protocol, increasing the percentage of events detected from less than 50% of the B-MAC to more than 90% when the traffic load is high (10s periodic interarrival time), for other loads the reliability is almost 100% in all the cases.

Table 2. Collective Reliability for 10s and 25s Periodic Interarrival Time

MAC Protocol	10s	25s
IEEE 802.11	0.998	0.999
B-MAC	0.490	0.997
LWT-MAC	0.986	0.999

6   Concluding Remarks

In this work a self-adaptive hybrid MAC protocol that combines unscheduled and scheduled accesses is presented. At low traffic loads the unscheduled access maintains a low energy consumption and when the load increases (for instance due to an event occurrence) the scheduled access provides small delay and higher throughput and reliability.

The wake up after transmissions allows to synchronize neighbouring nodes without the cost of creating, sharing and maintaining the schedule, apart from requiring lower synchronization capabilities if compared to the existing scheduled MAC protocols.

Results show that the LPL MAC protocol with scheduled wake up after transmissions is specially appealing for event-based WSNs as it can fit the requirements of periodic readings and react to sporadic changes in the network load.

Moreover, it significantly improves the collective QoS in terms of delay, bandwidth and reliability while keeping a low energy consumption.

References

1. Akyildiz, I.F., Su, W., Sankarasubramaniam, Y., Cayirci, E.: Wireless Sensor Networks: A Survey. Computer Networks 38(4), 393–422 (2002)
2. Kredo, K., Mohapatra, P.: Medium Access Control in Wireless Sensor Networks. Computer Networks 51(4), 961–994 (2007)
3. Chen, D., Varshney, P.K.: QoS Support in Wireless Sensor Networks: A Survey. In: Proceedings of the International Conference on Wireless Networks (ICWN 2004), pp. 227–233 (2004)
4. Lin, X.Z., Zhou, J.J., Mu, C.D.: Collective Real-Time QoS in Wireless Sensor Networks. In: Proceedings of the International Conference on Wireless Communications, Networking and Mobile Computing (WiCOM 2006), pp. 1–4 (2006)
5. IEEE Std 802.11. Wireless LAN Medium Access Control (MAC) and Physical Layer (PHY) Specifications. ANSI/IEEE Std 802.11, 1999 Edition (revised, 2003)
6. Ye, W., Heidemann, J., Estrin, D.: Medium Access Control with Coordinated Adaptive Sleeping for Wireless Sensor Networks. IEEE/ACM Transactions on Networking 12(3), 493–506 (2004)
7. IEEE Std 802.15.4. Wireless LAN Medium Access Control (MAC) and Physical Layer (PHY) Specifications for low-rate wireless personal area networks. IEEE Std 802.15.4, 2003 Edition (revised, 2006)
8. Polastre, J., Hill, J., Culler, D.: Versatile Low Power Media Access for Wireless Sensor Networks. In: Proceedings of the 2nd International Conference on Embedded Networked Sensor Systems, pp. 95–107. ACM Press, New York (2004)
9. Paek, K.J., Kim, J., Song, U.S., Hwang, C.S.: Priority-Based Medium Access Control Protocol for Providing QoS in Wireless Sensor Networks. IEICE Transactions on Information and Systems 90(9), 1448 (2007)
10. Liu, Y., Elhanany, I., Qi, H.: An Energy-efficient QoS-aware Media Access Control Protocol for Wireless Sensor Networks. In: Proceedings of the International Conference on Mobile Adhoc and Sensor Systems Conference, pp. 189–191 (2005)
11. Liu, Z., Elhanany, I.: RL-MAC: A QoS-Aware Reinforcement Learning based MAC Protocol for Wireless Sensor Networks. In: Proceedings of the IEEE International Conference on Networking, Sensing and Control (ICNSC 2006), pp. 768–773 (2006)
12. Chen, G., Branch, J., Pflug, M.J., Zhu, L., Szymanski, B.: SENSE: A Sensor Network Simulator. Advances in Pervasive Computing and Networking, pp. 249–269 (2004)
13. NS2. Network Simulator 2, http://www.isi.edu/nsnam/ns/

Implicit Sleep Mode Determination in Power Management of Event-Driven Deeply Embedded Systems

André Sieber, Karsten Walther, Stefan Nürnberger, and Jörg Nolte

Distributed Systems/Operating Systems group, BTU Cottbus
Konrad-Wachsmann-Allee 1
03046 Cottbus, Germany
{as,kwalther,snuernbe,jon}@informatik.tu-cottbus.de

Abstract. Energy consumption is a crucial factor for the lifetime of many embedded systems, especially wireless sensor networks. Most modern microcontrollers provide various low power sleep modes. Utilizing them can lead to great energy savings. In this paper we present an approach for power management in embedded systems, based on the event-driven operating system REFLEX. The implicit power management is mostly hardware independent, lightweight and efficiently chooses the optimal power saving mode of the microprocessor automatically.

Keywords: Embedded systems, energy, power management, sleep modes, sensor networks.

1 Introduction

Typical sensornet applications such as environmental monitoring demand that sensor nodes should work for months or even years with a single battery. Thus, saving energy is essential to archive this goal. Even in embedded systems with external power supply, saved energy helps to reduce costs. For minimal power consumption the appropriate hardware must be chosen. But the hardware must be appropriately utilized, too. This is a complex task that should not be the application programmer's burden. There is a wide range of low-power microcontrollers available, for example the Texas Instruments MSP 430 series[1]. These processors provide the programmer with fine grained control over components and sleep modes. To utilize these features the operating system should at least be able to provide the application with power saving mechanisms. It would be even better to do this implicitly without any need of control from the programmer.

Event-driven operating systems can potentially go into sleep mode if no event is pending and the hardware can wake up the rest of the system upon the occurrence of an event. Since most microcontrollers support a variety of sleep modes with different energy footprints, as shown for the MSP 430 in table 1, the selection of the mode can have an intense effect on the lifetime of battery powered devices. If the decision was wrong, the energy savings could be marginal or, even worse, events can get lost.

H. van den Berg et al. (Eds.): WWIC 2009, LNCS 5546, pp. 13–23, 2009.
© Springer-Verlag Berlin Heidelberg 2009

Table 1. Current consumption of MSP430 sleep modes @ 3V, 1MHz

mode	characteristics	consumption
active		$340\,\mu A$
lpm0	CLK and MCLK deactivated	$70\,\mu A$
lpm1	lpm0 + DCO deactivated	–
lpm2	lpm0 + SMCLK deactivated	$17\,\mu A$
lpm3	lpm1 + SMCLK deactivated	$1\,\mu A$
lpm4	all deactivated	$0.1\,\mu A$
	wakeup on external event only	

This paper is structured as follows. In section 2 the related work is presented. Section 3 gives a short overview of the REFLEX operating system. In section 4 the power management mechanisms of REFLEX are described. Early evaluation results are presented in section 5. Finally, a conclusion and future work are given in section 6.

2   Related Work

Operating systems for deeply embedded systems like sensor networks are becoming increasingly popular because they ease the use of the underlying hardware by introducing portability and providing standardised abstractions and interfaces. Their essential goal is to free the application programmer from issues like synchronisation, scheduling and most hardware dependencies, letting them concentrate on the application itself. The support for power management, especially computing the deepest possible sleep mode, vary from system to system.

The most common operating system for wireless sensor nodes is TinyOS[2]. It features a wide range of software modules and runs on various platforms. As it is an event-based operating system the scheduler of TinyOS puts the controller to sleep if the `task-queue` is empty and thus no work has to be done. TinyOS is capable of computing the deepest possible sleep mode autonomously. To do so, all components that change the state of the hardware and might influence the deepest possible sleep mode have to call the `McuPowerState.update()` function. This will set a `dirty-bit`, which is evaluated by the scheduler before entering sleep mode, resulting in a re-computation if set. The computation is done by reading all device registers. Because the update operation is executed atomically it can produce a significant overhead [3]. Additionally the function is hardware dependent and must be implemented for every platform. The `PowerOverride.lowestState()` function makes it possible for higher level components to influence the chosen sleep mode. This could be useful for components which have requirements that can not be described with the device registers, e.g. wakeup latencies.

Other event-driven sensor node operating systems like SOS[4] and Retos[5] leave the power management to the programmer and do not implement any deepest sleep mode computation. Contiki[6] can not take advantages from any sleep mode because of its polling methodology for interrupt handlers.

In thread based operating systems for deeply embedded systems it is more challenging to determine the possibility of going into a certain sleep mode. The scheduler has to determine if all threads are blocked for some reason (e.g. waiting on I/O) or are idle. Systems like AVRx [10] or freeRTOS [11] have no support for power management and leave sleep mode implementation to the programmer. Mantis [9] has support for sleep modes. It provides a `mos_thread_sleep()` function, similar to the UNIX sleep(). Every thread has to call it with the desired duration of the sleep period. Mantis only distinguishes between an idle sleep mode and a deeper sleep mode. The first is used if the system waits on IO, the second if all threads have called the `mos_thread_sleep()` function.

As shown above, most operating systems for deeply embedded systems provide no support for power management, regardless of they are event or thread based. Systems which provide management features suffer from hardware dependencies and overhead or need advanced attention of the application programmer.

3   Reflex

REFLEX (**R**eal-time **E**vent **FL**ow **EX**ecutive) is an operating system implemented in C++, targeting deeply embedded systems and wireless sensor nodes. It is based on the so called event flow model presented in [7]. In REFLEX so called activities are schedulable entities, which are triggered if something was posted to their associated event buffers. Initial source of all activity in the system are interrupts.

Reflex features a scheduling framework which allows a change of strategy with a few lines of code. There are various schemes implemented, e.g. Earliest Deadline First, First Come First Serve, Time Triggered and Fixed Priority. Although reflex is a single stack system with run to completion semantics, it is possible to use preemptive scheduling, since all preemptive schemes can be mapped to a Last In First Out order.

Logical correlated activities and interrupt handlers are grouped to components, connected among each other with event channels. These channels can

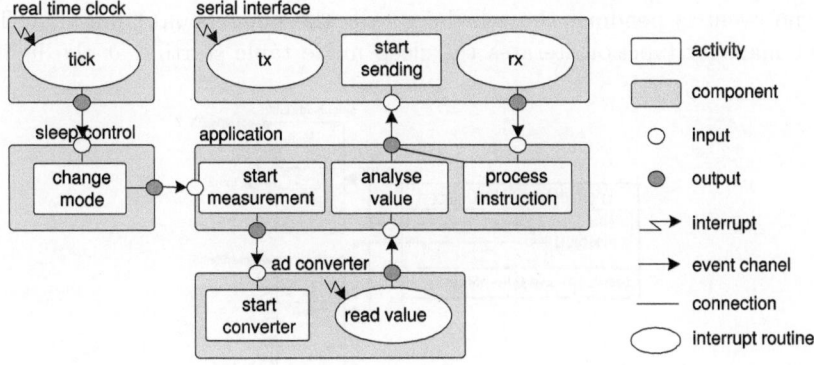

Fig. 1. REFLEX application example

have pre computing abilities, e.g. filters which pass only events with a certain property, dividers which pass only the n-th event etc.

Figure 1 shows a simple example containing the elements of a common RE-FLEX application. The interrupt handler in the timer component writes data into the event buffer of the AD Converter component using an event channel, which forces the activity of the converter to sample a value. When the value is available the ADC interrupt arises and is handled by writing the value to the event buffer of the application logic component. The associated activity implements the threshold logic which sends the data via the serial interface component if it is above a certain threshold.

4   Power Management in Reflex

The power management in REFLEX divides two abstractions, a system view and a user view. The system view is responsible for the determination of the deepest possible sleep mode. The user view provides the programmer with two instruments, namely groups and modes, to ease the handling of all hard and software components.

4.1   System View

The system provides the class `EnergyManageAble`, each class derived from it is concerned by the power management. Manageable objects can be interrupt handlers as well as activities or event channels.

Figure 2 shows how the system view is used to determine the deepest possible sleep mode. Every instance of a component has a variable which specifies the deepest possible sleep mode that may be used when it is active. This value is assigned during initialization for non hardware specific objects or hard coded for hardware specific objects. The power manager contains a table with counters for every available sleep mode of the microcontroller used. If a component is enabled it signals its deepest possible sleep mode to the power manager by increasing the corresponding counter in the sleep mode table. If a component is disabled, the counter of its sleep mode is decreased.

If no event is pending, the scheduler calls the `powerDown()` function. This power manager function iterates the sleep mode table starting at the lightest

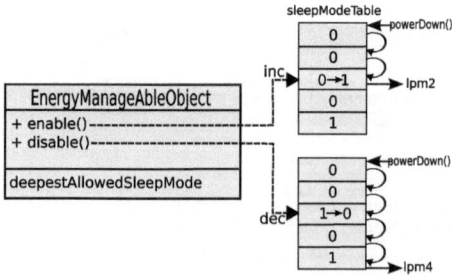

Fig. 2. REFLEX power management view

```
0   //PowerManager.h

    uint8 useCounts[NrOfSleepModes];
    // array of functionpointers the the platform dependent
        sleepfunctions
    static void (* const modes[NrOfSleepModes])();
5
    void reflex::PowerManager::powerDown() const
    {
      int i = 0;
      //look for deepest possible powerdown method
10    while(!useCounts[i] && i<(NrOfSleepModes-1))
      {
        i++;
      }
      (*modes[i])();//call powerdown method
15  }
```

Fig. 3. Power management table implementation

mode. The first value different from zero is the deepest possible sleep mode. To guarantee this, the table is initialized with zero for all counters except the deepest mode which is initialized to one. Figure 3 shows the hardware independent implementation of the **powerDown()** function.

In contrast to TinyOS, it is not necessary to evaluate the complete machine state, which makes the changing of the lowest possible sleep mode very lightweight.

The sleep mode counter table is the only hardware specific part of the power manager, since it can have a different size depending on the microcontroller used. Figure 4 shows the implementation of the sleep mode table for the MSP 430. The table contains pointers to the assembler functions which enable the interrupts and send the controller to the specific sleep mode.

A state override is possible to prevent the system to go below a certain sleep mode or to sleep at all. This is possible by using an energy manageable object which increments the counter in the power management table for a certain wanted mode. This may be useful for reducing wakeup times.

Since the initial source for activity are the interrupts, they define the deepest possible sleep mode. In general there are two types of interrupts, primary and secondary. The first are caused by external events, the second are a result of software events. E.g. the TX interrupt of a serial connection has only to be active when a send operation is in progress, if it is finished, the driver of the serial connection can deactivate the interrupt and possibly change the deepest possible sleep mode. Thus the power management approach is implicit for secondary interrupts, because the drivers know when a interrupt has to be enabled.

Primary interrupts can not be deactivated implicitly. Their state is determined by the current stage of the application. Sampling of sensors and sleeping for

```
0  //ControllerPowerManagement.cc
   extern "C" {
   void _lpm0();
   void _lpm1();
   void _lpm2();
5  void _lpm3();
   void _lpm4();
   }

   void (* const PowerManager::modes[NrOfSleepModes])()={_lpm0,
       _lpm1, _lpm2, _lpm3, _lpm4};
```

Fig. 4. Sleep mode implementation

a given time needs different interrupts at different times. The decision which interrupt has to be active must be decided by the application programmer.

4.2   User View

The user view provides two mechanisms to ease the control of hardware and even software components, groups and modes.

At startup, each manageable object is registered with the power management and assigned to one or more programmer defined groups as shown in figure 5.

During operation, groups can be independently activated and deactivated. This allows to easily activate or deactivate any number of objects with only one method call. If a manageable object is member of multiple groups it is only deactivated when all of these groups are deactivated.

Modes are defined to switch easily between active groups and thus utilizing different hardware configurations. This makes it possible to divide the execution of an application into different phases, while ensuring that the hardware components are active when demanded. The programmer is responsible for changing the modes. For example a timer driven module can be used for mode changes.

```
0  //set group memberships
   timer.setGroups( SLEEP | AWAKE );
   serial.RX.setGroups( AWAKE );
   sensor.setGroups( AWAKE );

5  //activate all members from group sleep
   powerManager.enableGroup(SLEEP)

   //switch mode from group SLEEP to AWAKE
   powerManager.switchMode(AWAKE,SLEEP);
```

Fig. 5. Groups and Modes example

4.3   Example

Figure 6 shows a simple power managed sample application. It consists of a driver for the timer, a communication interface and the application logic itself. Periodically the application wakes up and waits for commands from the interface. When demanded sensor data are send over the same interface. The application has two modes, `asleep` and `awake`. The timer is active in both, the receiving part of the communication interface only in the second one. Part a) of the figure shows the `asleep` mode with the corresponding sleep mode table entries. Since the timer is assumed to be active in the dream mode of the processor, this is the deepest possible sleep mode. When the timer expires, the change mode activity within the application component is scheduled, which is responsible for resetting the timer and switching between `asleep` and `awake`. Due to the switch to the `awake` mode, the RX part of the interface is enabled, which leads to a modification of the sleep mode table and thus the deepest possible sleep mode as shown in part b) of the figure.

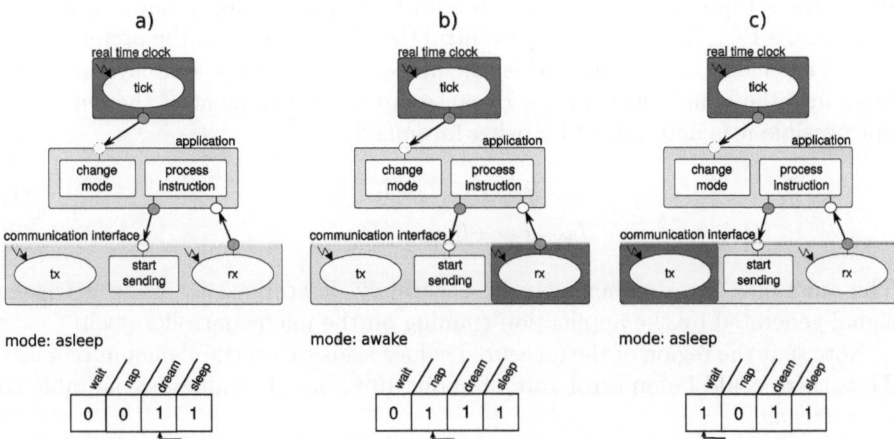

Fig. 6. Power management example showing the influence of drivers to the deepest possible sleep mode

In the last part of the figure, data is sent via the interface. As a secondary interrupt the driver itself is responsible to enable the device at request and thus change the sleep mode table. Although a change to the `asleep` mode can occur, the send operation will finished, because the interface driver sets the sleep mode needed by the device.

5   Evaluation

To show the efficiency and comparability of the presented power management scheme, we devised a simple experiment. In the next section we first explain the experimental setup. After that the results will be discussed.

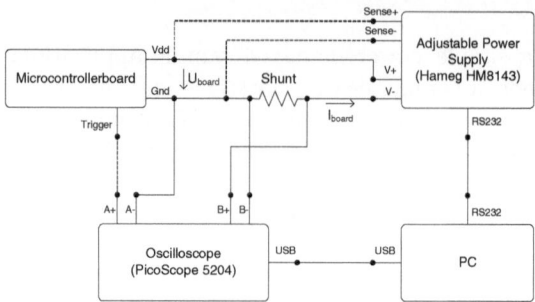

Fig. 7. Measurement Setup

5.1   Experimental Setup

The measurement setup is shown in figure 7. Besides the device under measurement, the setup contains a shunt, an adjustable power supply unit, an oscilloscope and a PC. The PC is used to control the oscilloscope and the power supply unit. The energy consumption P is calculated by the voltage and current which flows into the board (formula 1). Because direct measurement of the current is not possible it is determined by using formula 2.

$$P = U_{board} * I_{board} dt \tag{1}$$

$$I_{board} = -U_{shunt} / R_{shunt} \tag{2}$$

The start of a measurement can be accurately determined by using a trigger signal generated by the application running on the microcontroller itself.

Note that the region of the measured values leads to a certain amount of noise. Thus there could be an error rate of about 10%, but the values are suitable to show trends.

The measurements where performed using a TMoteSky[8] equipped with a Texas Instruments MSP 430 series microcontroller, which has special hardware support for low-power applications e.g. detailed sleep modes and fast wakeup times.

5.2   Results

We first used a simple evaluation application, which sends data over the serial connection every second. This application was chosen because it shows the characteristics of a typical sensornet application, such as a low duty cycle and periodically sending data over an interface. In figure 8 and table 2 the power consumption for different versions of the power management schemes in REFLEX are compared.

With deactivated power management, there is no difference in power consumption regardless of whether the serial port is active or not. In contrast, deactivating the idle core (the highest possible sleep mode called lpm0) makes

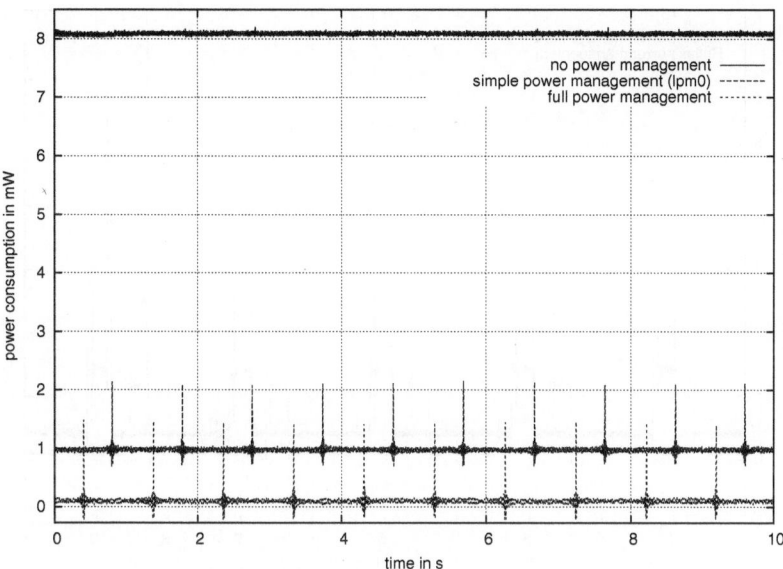

Fig. 8. Power consumption of REFLEX power management schemes at 1MHz and 3V

the send activity clearly identifiable and shows a significantly amount of energy savings. This simple power management reduces the energy consumption to about 11% of the unmanaged version. Using the power management approach presented in section 4 consequently to utilize the deepest possible sleep mode reduces the energy consumption to 1%.

For comparison purpose TinyOS 2.0.2 was used. The application was a simple sense and send application utilizing the implicit power management of both systems. It samples the SHT11 temperature and humidity sensor of the TMoteSky every two seconds, aggregates these values four times and sends them over the serial connection at the beginning of the fifth phase.

The results are shown in figure 9 and table 3.

For the given application the results show that the power management of REFLEX is more effective than that of TinyOS. The REFLEX application consumes about 38% respectively 51% less energy, depending on the voltage. The main reason is that the power consumption during sleep is significantly higher in TinyOS than in REFLEX. This is caused by the baudrate generator which is used by the driver for the serial connection. Since this interface was only used unidirectional, it should be deactivated by the driver when there is no data do

Table 2. Average energy consumption of REFLEX in mW per Second

	w/o	lpm0	full
TMoteSky 2,2V @ 1MHz	3.072	0.418	0.039
TMoteSky 3V @ 1MHz	8.263	0.889	0.083

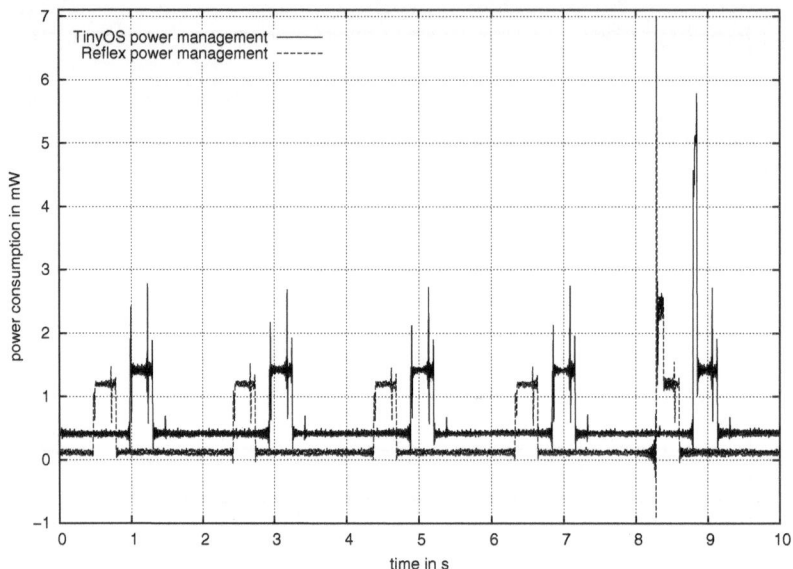

Fig. 9. Power consumption of TinyOS and REFLEX at 1MHz and 3V

Table 3. Average energy consumption of TinyOS and REFLEX in mW per second

	Reflex	TinyOS
TMoteSky 2.2V @ 1MHz	0.193	0.316
TMoteSky 3V @ 1MHz	0.281	0.575

transmit. This was accounted for by REFLEX but not by TinyOS, so the generator runs all the time consuming energy. At the beginning of the send operation there is a higher spike in the REFLEX curve due to the necessary startup of the baudrate generator.

6   Conclusion and Future Work

We presented an efficient way of determining the deepest possible sleep mode in event driven systems and its implementation in the REFLEX operation system. The presented power management is lightweight and hardware independent. Due to the event flow model and its implementation REFLEX is capable of implicitly deactivating parts of the hardware which are not used. This proves to be an advantage as the results show. Further examination of additional applications utilizing more hardware components is needed to confirm the results.

One way to extend the power management is to ease the handling of the user view and free the user of any power management related tasks. Another possible way is to integrate dynamic frequency and/or voltage scaling. In [12] the authors propose a combination of sleep modes and dynamic frequency scaling, for applications with low workload, like in sensor networks. For applications with significant workloads dynamic voltage and frequency scaling savings exceed the costs of a switchable power supply. Since most microcontrollers provide a simple way to change frequency on the fly, it could be worthwhile to integrate dynamic frequency scaling into the power management, for example by assigning frequencies to groups.

Acknowledgments

This work was partially supported within the TANDEM project by the Inno-Profile program of the German Federal Ministry of Education and Research.

References

1. Texas Instruments, MSP430 series Datasheet, http://www.msp430.com
2. Hill, J., Szewczyk, R., Woo, A., Hollar, S., Culler, D., Pister, K.: System Architecture Directions for Networked Sensors. In: The 9th International Conference on Architectural Support for Programming Languages and Operating Systems (ASPLOS-IX) (2000)
3. Szewczyk, R., Levis, P., Turon, M., Nachman, L., Buonadonna, P., Handziski, V.: TinyOS Microcontroller Power Management Documentation TEP112, http://www.tinyos.net/tinyos-2.x/doc/html/tep112.html
4. Han, C.-C., Ram Kumar, R.S., Kohler, E., Srivastava, M.: A dynamic operatingsystem for sensor nodes. In: Proc. of the 3rd international conference on Mobile systems, applications, and services MobiSys (2005)
5. Cha, H., Choi, S., Jung, I., Kim, H., Shin, H., Yoo, J., Yoon, C.: Retos: resilient, expandable, and threaded operating system for wireless sensor networks. In: Proc. of the 6th intl. conf. on Information processing in sensor networks (IPSN 2007) (2007)
6. Dunkels, A., Gronvall, B., Voigt, T.: Contiki - a lightweight and exible operating system for tiny networked sensors. In: Proceedings of the First IEEE Workshop on Embedded Networked Sensors (2004)
7. Walther, K., Nolte, J.: A flexible scheduling framework for deeply embedded systems. In: Proc. of 4th IEEE International Symposium on Embedded Computing (2007)
8. Moteiv Corperation, TMote Sky Datasheet (2006), http://www.sentilla.com/moteivtransition.html
9. Bhatti, S., Carlson, J., Dai, H., Deng, J., Rose, J., Sheth, A., Shucker, B., Gruenwald, C., Torgerson, A., Han, R.: MANTIS OS: An Embedded Multithreaded Operating System for Wireless Micro Sensor Platforms. In: ACM/Kluwer Mobile Networks & Applications (MONET), Special Issue on Wireless Sensor Networks (2005)
10. AVRX, http://www.barello.net/avrx
11. FreeRTOS, http://www.freertos.org/
12. Ghattas, R., Dean, A.G.: Energy management for commodity short-bit-width microcontrollers. In: Proceedings of the 2005 international Conference on Compilers, Architectures and Synthesis For Embedded Systems (CASES 2005) (2005)

On Prolonging Sensornode Gateway Lifetime by Adapting Its Duty Cycle

Marcin Brzozowski and Peter Langendoerfer

IHP GmbH
Im Technologiepark 25
15236 Frankfurt (Oder)
Germany
{brzozowski,langendoerfer}@ihp-microelectronics.com

Abstract. In this paper we discuss the lifetime of an battery powered device which acts as a gateway between a wireless sensor network and a standard network. The wireless communication standards used are IEEE802.15.4 and IEEE802.11g. The two parameters which are of utmost importance but also contradicting each other are communication delay and lifetime of the gateway node. Our results clearly show that duty cycling i.e. switching off the gateway node improves its lifetime. Depending on the radio modules used the lifetime can be increased from seven hours up to 3 months using only 3 AA batteries. Our results also proove that prolonging the sleep intervals beyond a certain limit (about 10 seconds for a typical WLAN PC card) does not longer improve the lifetime but only worsens the delay.

Keywords: Sensor networks, clustering, gateway, low duty cycle, MAC, 802.11g, 802.15.4.

1 Introduction

To connect wireless sensor networks to the outside world intermediate systems providing gateway functionality are used [11,7,3,8,4]. Often base stations are considered to be these systems. In order to improve the scalability of WSNs we assume that battery powered clusterheads within the WSN are equipped with two different radio modules and provide the gateway functionality. For such a special sensor node the uptime is of utmost importance. The second important aspect which has a significant influence on the usability of such a WSN is the delay of messages sent into and out of the WSN respectively.

The main contribution of this work is the evidence that the proper choice of the sleep period does not affect an event delay report significantly, and prolongs the gateway's lifetime. We showed that a clusterhead connecting 802.15.4 sensor network to WLAN network can work for almost 2 months with only 3 AA batteries. The event report delay is 3 seconds on average, and 6 seconds in the worst case. In addition, our analysis revealed that a duty cycle approach cannot extend the lifetime of the gateway node across a certain time (3 months).

H. van den Berg et al. (Eds.): WWIC 2009, LNCS 5546, pp. 24–35, 2009.

I.e. extending the sleep period over certain limit does not prolong the lifetime of the clusterhead considerably but increases the event report delay unnecessarily!

The paper is organized as follows: Section 2 introduces related work. The network architecture we assumed for our analysis as well as the information flow are presented in Section 3. Our theoretical analysis is described in Section 4. Section 5 concludes the paper.

2 Related Work

Several research efforts addressed the low duty cycle and rendezvous problem, that is, the need of the sender and the receiver to be awake at the same time.

Ref. [10] classifies rendezvous schemes into three categories: purely synchronous (nodes agree on the next communication time), purely asynchronous (a node wakes up another node, e.g., wake-up radio) and pseudo-asynchronous (nodes use an underlying periodic wake-up scheme).

SMAC[17] is an example of the purely synchronous scheme. The exact time of sleep and active state determines the schedule of the node. As neighboring nodes coordinate their schedule, they are active at the same time.

STEM (Sparse Topology and Energy Management)[16] and preamble sampling [5] are examples of the pseudo-asynchronous scheme. A transmitter sends several requests, before it sends data. A potential receiver wakes up periodically and monitors the channel for a short period for incoming transmission requests. In the preamble sampling[5] technique, a transmitter sends a long preamble in front of every message. WiseMAC[1] extends the preamble sampling [5] by learning the neighbors' sampling schedules, and reducing the need of long preambles.

Several research efforts [11,7,3,8,4] analysed the use of gateway between a sensor network and another network. Ref. [11,4,8] discusses the connection of a sensor network with WLAN interface. Moreover, the authors [4] presents the connection of a 802.15.4 sensor network with WLAN, as we do in this paper. General ideas of interoperability between a sensor network and a TCP/IP network shows [3].

3 Architecture and Protocol

3.1 Architecture

We consider a sensor network which consists of three device types: a base station, a clusterhead and sensor nodes. The base station has WLAN transceiver, which is always powered on. The clusterhead works as a gateway between the sensor network and the base station. The clusterhead has two radio modules: IEEE 802.15.4 and IEEE802.11g. To communicate with sensor nodes, the clusterhead uses the transceiver based on IEEE 802.15.4 standard. The clusterhead sends data to the base station and receives data from it using WLAN. The clusterhead alternates between two states: active (sending, listening, receiving) and sleep. Each sensor node has an 802.15.4 transceiver. To prolong the lifetime, each sensor node keeps the transceiver powered off most of the time. Figure 1 illustrates such an architecture.

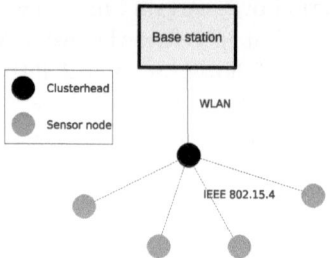

Fig. 1. Architecture

3.2   Communication Flows

There are two types of communication flows: upward and downward. In upward communication, sensor nodes send data to the base station via the clusterhead. As soon as a sensor node detects an event, it sends an event message to the clusterhead. After reception of the message, the clusterhead forwards it to the base station. To find out whether sensor nodes are still working, each sensor node sends periodically a message to the clusterhead, which forwards it to the base station. If the base station does not receive messages from a sensor node for some time, the base station assumes the node does not work. In downward communication, the base station sends occasionally data to sensor nodes via the clusterhead. Although the protocol supports communication in both directions, we use the downward communication rarely (e.g. to update the code on sensor nodes, to send new data queries).

3.3   Problem Statement

To prolong the clusterhead's lifetime, we use a duty cycle protocol. Such a protocol puts the clusterhead in the sleep state whenever possible, allowing a longer lifetime. However, when the clusterhead is in the sleep state, it cannot forward any event reports from sensor nodes to the base station. Event reports flow from a sensor node to the base station only when the clusterhead is active, as depicted in Figure 2. Thus, the duty cycle protocol of the clusterhead influences the event report delay: the longer the sleep period, the greater the event report delay. There are two extremes in the duty cycle protocol of the clusterhead, which provide the smallest possible delay option 1 and the longest lifetime option 2:

1. The clusterhead is always on. As soon as a sensor node detects an event, it sends data to the base station via the clusterhead. In that case, there is the minimum event report delay. However, the clusterhead consumes the whole energy quickly (e.g. tmote sky[12] equipped with OWLAN211g[2] transceiver and 3 AA batteries consumes the whole energy in a few hours).
2. The clusterhead sleeps only, and achieves the longest possible lifetime (almost 3 months for above-said tmote sky node). As the clusterhead sleeps only, it

cannot forward data between two networks. Therefore, no data flows between two networks, and sensor nodes cannot report any events.

In this paper, we focus on the following problem: how to prolong the clusterhead's lifetime and do not worsen the event report delay significantly. Therefore, we examine trade-off between the clusterhead's lifetime and the event report delay.

3.4   Duty Cycle MAC

We propose a pseudo-asynchronous duty cycle protocol, which can be realized using the 802.15.4 guaranteed time slot (GTS) feature. The clusterhead sends periodically a beacon to sensor nodes using the 802.15.4 transceiver. After sending a beacon, the clusterhead listens for possible transmissions from sensor nodes. Similarly, the clusterhead sends a beacon to the base station with WLAN, and listens shortly for incoming data from the station. Figure 2 shows the principle of our duty cycle protocol. In the following subsections we present details on how nodes learn the beacon period, how the clock drift is countered and how normal communication is done.

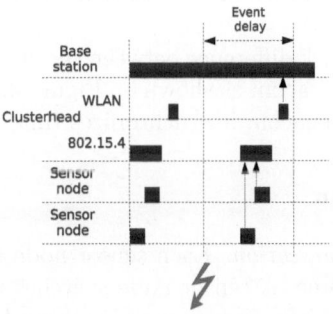

Fig. 2. A duty cycle protocol for the clusterhead and sensor nodes. The bar indicates active times in which data can be received and sent.

3.4.1   Joining the Network
After a node is placed in the network, it is not aware of the clusterhead active times. Therefore, the node cannot send any data to the clusterhead. To join the network, the node listens, in the worst case, for the whole beacon period. After receiving a beacon, it sends a message to the clusterhead and announces its presence. The clusterhead registers the new node, and assigns a new time slot. From now on, the node can send data to the clusterhead.

3.4.2   Clock Drift
As all the sensor nodes are aware of the beacon period, they can calculate the next beacon time of the clusterhead. A sensor node estimates the next beacon time of the clusterhead by adding the beacon period to the reception time of the last beacon.

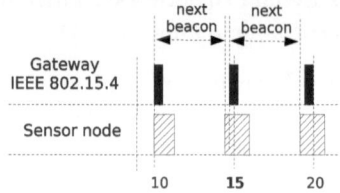

Fig. 3. Each sensor node wakes up earlier by the guard time to compensate clock drift

Because of the clock drift, a sensor node may wake up too late or too soon to receive the beacon of the clusterhead. Thus, sensor node may miss a beacon.

To counter the clock drift problem, sensor nodes use guard times. As the nodes are aware of their clock inaccuracy, they estimate the worst possible clock drift for the beacon period, referred to as guard time. Each sensor node wakes up not exactly after the beacon period of the clusterhead, but earlier by the guard time. For instance, the clock drift of tmote sky clock is 40 ppm. In the worst case, the time difference between two nodes in a second is about 80 microseconds. For the beacon period of 5 seconds, a sensor node wakes up earlier by 4 ms to compensate clock drift.

In this protocol, the clock difference between the clusterhead and a sensor node arises only from the last beacon, as shown on Figure 3. In other words, each time a sensor node receives a beacon, it synchronizes time to the clusterhead.

3.4.3   Communication

3.4.3.1   Upward Communication. Each sensor node keeps its transceiver powered down most of the time. A sensor node switches the transceiver on only to receive a beacon. After a beacon reception, each node sends event information to the clusterhead, if it detected an event recently. If the node did not detect any event, it sends only an "alive" message. To avoid collisions, which use the 802.15.4 guaranteed time slot (GTS) feature, as already mentioned. Each node is allowed to send data only during its time slot. Nodes get their time slots from the clusterhead after powering on and joining the network.

The clusterhead can receive data only for a short time after sending a beacon. Therefore, after an event occurred, a sensor node cannot send data to the clusterhead immediately. It waits until the clusterhead sends a beacon and is ready to receive data. Obviously, this involves an event report delay. In the worst case, the node waits for almost the whole beacon period.

3.4.3.2   Downward Communication. As the clusterhead keeps its WLAN transceiver powered off most of the time, the base station cannot send data to the clusterhead any time. When the clusterhead switches WLAN module on, it sends a message to base station. After receiving the message, the base station sends data to the clusterhead, which forwards it to sensor nodes by piggybacking it into beacons.

4   Evaluation

4.1   Model

We estimate the lifetime of the clusterhead in days as follows:

$$Lifetime = \frac{Q}{E_{day}} \tag{1}$$

where

- Q is available energy [mAh]
- E_{day} is the total daily energy consumption of the clusterhead [mAh/day].

We estimate the energy consumption of a single transceiver as the sum of the energy consumed during active (sending, receiving) and sleep states. As the transceiver is active only when sending beacons, and receiving data thereafter, we determine the energy consumption in active state in relation to beacons. Thus, we estimate the total daily energy consumption as follows:

$$\begin{aligned}
E_{day} = E_{tx} + E_{rx} + E_{sleep} = \\
B \cdot (t_{startup}I_{startup} + t_{tx}I_{tx} + \\
t_{rx}I_{rx} + t_{shutdown}I_{shutdown}) + \\
T_{sleep}I_{sleep}
\end{aligned} \tag{2}$$

where:

- E_{tx}, E_{rx}, E_{sleep} is the daily energy consumption when sending, receiving and sleeping[1]
- I_{tx}, I_{rx}, I_{sleep}, $I_{startup}$, $I_{shutdown}$ is the (average) current consumption when sending, receiving, sleeping, starting up and shutting down the transceiver respectively
- t_{tx}, t_{rx} are the times of sending a single beacon and listening for incoming data thereafter
- $t_{startup}, t_{shutdown}$ are the times to start up and shutdown the transceiver
- T_{sleep} is the total sleep time of the node a day
- B is the number of beacons a day

The clusterhead sends and receives data periodically, after the beacon period. Knowing the transceiver data rate and the amount of data transmitted, we can estimate the single transmit (t_{tx}) and receive(t_{rx}) times.

The clusterhead sleeps almost all the time, apart from when transmitting and receiving. Thus, we determine the sleep time a day by subtracting the transmit and receive times:

$$T_{sleep} = T_{day} - B \cdot (t_{startup} + t_{tx} + t_{rx} + t_{shutdown}) \tag{3}$$

---

[1] The total energy consumed by the transceiver when sending, receiving and sleeping contains also the energy consumed by the microcontroller in these states.

where T_{day} is the number of time units (e.g. seconds) a day; the transmit and receive times must be specified in the same unit. For example, to calculate the total sleep time a day in seconds, T_{day} equals 86 400 (seconds a day).

We estimate the number of beacons a day B:

$$B = \frac{T_{day}}{T_{beacon}} \qquad (4)$$

where T_{beacon} is the beacon period, and must be expressed in the same time unit as T_{day}.

Generally, we estimate the daily energy consumption of the clusterhead having n transceivers as follows:

$$
\begin{aligned}
E_{day} = T_{day} \sum_{i=1}^{n} \{ &\frac{1}{T_{beacon\_i}} \cdot \\
&[t_{startup\_i}(I_{startup\_i} - I_{sleep\_i}) + t_{tx\_i}(I_{tx\_i} - I_{sleep\_i}) + \\
&t_{rx\_i}(I_{rx\_i} - I_{sleep\_i}) + t_{shutdown\_i}(I_{shutdown\_i} - I_{shutdown\_i})] + \\
&T_{day}I_{sleep\_i}\}
\end{aligned}
\qquad (5)
$$

In the presented model, we do not consider energy consumed by data retransmission protocols, like ARQ (Automatic Repeat reQuest) protocol[6] .

4.2   Scenario

In our scenario the clusterhead controls 10 sensor nodes. The number of sensor nodes influences the listening time of the clusterhead: the more sensor nodes, the longer the clusterhead listens for incoming data.

In this scenario, each sensor node sends 128 bytes of data after receiving a beacon.

In our analysis we used tmote sky[12] sensor nodes for both the clusterhead and sensor nodes. The current consumption values are taken from tmote datasheet[12], and from measurements[9]. We present these values in Table 1.

As tmote sky has only a 802.15.4 transceiver, we must use an additional WLAN transceiver for the clusterhead. In our analysis we used two WLAN modules. The first module, Orinoco 11b PC Card[14], is a standard wireless PCMCIA card. The second, OWLAN211g[2], is a WLAN transceiver designed for low power devices. We present the current consumption of these two WLAN

Table 1. Current consumption of tmote sky

Send, MCU On	20 mA
Receive, MCU On	21 mA
Radio Off, MCU On	2 mA
Radio Off, MCU Off	0.01 mA

Table 2. Current consumption of WLAN transceivers

	Orinoco 11b	OWLAN211g
Tx	285 mA	170 mA
Rx	185 mA	170 mA
Sleep	9 mA	0.8 mA

modules in Table 2. OWLAN211g[2] datasheet does not mention the receive energy consumption. Thus, we assumed that the receive energy consumption equals the transmit energy consumption.

We equipped the clusterhead with three rechargeable Sanyo eneloop[15] batteries. We decided to use these batteries, since around 90% of their capacity can be effectively used. Sanyo eneloops deliver 90% of the capacity at over 1.2V[15], and the three batteries connected in-series deliver 3.6V, which is sufficient for 802.15.4 transceiver of tmote sky, and for OWLAN211g module. However, Orinoco 11b PC requires 5V, and needs additional power supply for a real world application.

The self-discharge rate of these batteries is only about 15% a year. Although there are non-rechargeable batteries with a better self-discharge rate, namely few percent a year, most of them have a linear discharge rate[13]. Hence, these batteries deliver the current for a longer time than eneloops, but at the voltage lower than 1.2V, which is insufficient.

4.3   Lifetime and Event Delay Trade-Off

In order to determine the lifetime of the sensor nodes and of the clusterhead we applied the equations 1 and 5. We applied the energy consumption of 802.15.4 and WLAN transceivers as shown in table 2. We omitted the energy consumed during startup and shutdown of the transceivers, as it is relatively small in low duty cycle protocols (Figure 5 reveals that the clusterhead consumes plenty of energy in sleep mode).

Using the scenario described above, the sensor nodes work for almost two years for a beacon period of 5 seconds, and 3 years for 10 seconds beacon period. We focus on the analysis of the lifetime of the clusterhead, and on the event report delay.

We examined the relationship between the event report delay and the clusterhead's lifetime. The beacon period of the clusterhead determines the event report delay. In the worst case, a sensor node waits almost the whole beacon period before sending data. Thus, the event report delay equals to the beacon period in that case. However, on average the event report delay equals to the half of the beacon period.

As the clusterhead, we analyzed tmote sky sensor node, which has 802.15.4 transceiver. Additionaly, we fitted the clusterhead with two different WLAN transceivers: Orinoco and OWLAN211g. Figure 4 presents the results.

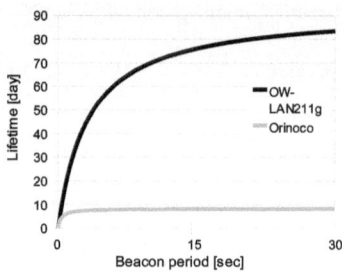

Fig. 4. The lifetime of the clusterhead for two different WLAN transceivers

The clusterhead equipped with OWLAN211g module achieves longer lifetime than the clusterhead with Orinoco transceiver, as expected. In the former case, the clusterhead works for more than 80 days, if it sleeps all the time. In the latter case, the clusterhead dies after less than 10 days. The difference in lifetimes results mostly from the energy consumption in the WLAN sleep state: Orinoco consumes about 10 times more energy in sleep state than OWLAN211, as presented in Table 2.

We prolong the lifetime of the clusterhead by increasing the beacon period, and the event report delay consequently. The clusterhead sends fewer beacons, and listens for incoming data less frequently. However, from a certain point, any increase in the beacon period does not prolong the lifetime significantly! For example, the change of the beacon period by 4 seconds results in various lifetime prolongations. Extending the beacon period from 1 second to 5 seconds, for the clusterhead using OWLAN211g, prolongs the lifetime by more than 30 days. However, the beacon period change from 11 seconds to 15 seconds results in only 5 day prolongation. With the change from 21 to 25 seconds, the clusterhead works longer only by less than 2 days. These observations reveal that too long beacon periods do not prolong the lifetime significantly.

We analyzed the energy consumption of transceivers, 802.15.4 and WLAN (Orinoco), in active and sleep modes. Figure 5 shows the results. Obviously, when we prolong the beacon period, the clusterhead stays longer in the sleep mode, and therefore consumes less energy. For example, the clusterhead consumes 600 mAh a day for the beacon period equal to 200 ms. The clusterhead consumes 223 mAh a day, when it sends beacons every 10 seconds.

However, the longer the beacon period is the more energy the WLAN transceiver consumes in the sleep mode. For example, the WLAN sleep mode consumes around 34 per cent (205 mAh a day) of the whole energy (600 mAh a day), when the clusterhead sends beacons with the 200 msec period. For the beacon period equal to 30 seconds, the WLAN sleep mode consumes 99 per cent (216 mAh a day) of the whole energy (219 mAh a day). As the WLAN sleep becomes the major energy consumer, any increase in beacon period does not lower the energy consumption significantly, and does not prolong the lifetime. For that reason, from a certain beacon period, we cannot prolong the lifetime

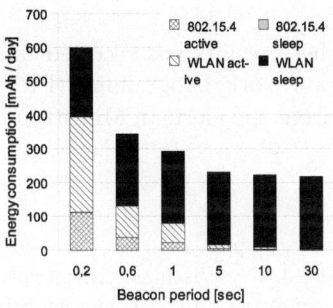

Fig. 5. Energy consumption of both clusterhead's transceivers, 802.15.4 and WLAN (Orinoco), for various beacon periods

significantly by increasing the beacon period. Therefore, any increase in the beacon period does not affect the lifetime, but only increases the event report delay unnecessarily.

5   Conclusion

Duty cycling of sensor nodes is a well known technique to increas their lifetime. It is also obvious that this approach comes at cost of resynchronization of sensor nodes, and of increased transmission delay. In this paper we have introduces an analytical model to investigate the relationship between duty cycles and transmission delay. In this paper we have analysed the liftime of a battery powered gateway node since it is the key position in a WSN. Our key findings are:

1. duty cycling helps to increase the lifetime of the sensor node significantly, in our example from several hours to up to 3 month.
2. duty cycling has some influence on the transmission delay, but it can be kept reasonably small; we achieved 6 seconds transmission delay, in the worst case, and the lifetime of 2 months.
3. extending the sleep interval above a certain limit does no longer improve the lifetime, but increases only the transmission delay; in our example this limit was about 2 minutes for a low power WLAN module, and 10 seconds for a typical WLAN card.
4. the quality of the sleep modi of the tranceiver modules is one of the key parameters when applying duty cycling.

Our analytical model can be used to engineer the sleep periods of wireless sensor networks in very early stages. The benefit is that in addition to the lifetime the transmission delay can be taken into account. In our future work we will integrate means to express transmission failures in order to retrieve results even more close to the real world behaviour when estimating delay times and lifetime of nodes.

Acknowledgment

The research leading to these results has received funding from the European Community's Seventh Framework Programme (FP7/2007-2013) under grant agreement n° 225186 and from the German Ministry of Education and Research under grant n° 03IP601.

References

1. El-Hoiydi, A., Decotignie, J.-D.: Wisemac: An ultra low power mac protocol for multi-hop wireless sensor networks. In: Nikoletseas, S.E., Rolim, J.D.P. (eds.) AL-GOSENSORS 2004. LNCS, vol. 3121, pp. 18–31. Springer, Heidelberg (2004)
2. connectBlue. Product Brief OWLAN211g, http://www.connectblue.com
3. Dunkels, A., Alonso, J., Voigt, T., Ritter, H., Schiller, J.H.: Connecting wireless sensornets with TCP/IP networks. In: Langendoerfer, P., Liu, M., Matta, I., Tsaoussidis, V. (eds.) WWIC 2004. LNCS, vol. 2957, pp. 143–152. Springer, Heidelberg (2004)
4. Dutta, P., Hui, J., Jeong, J., Kim, S., Sharp, C., Taneja, J., Tolle, G., Whitehouse, K., Culler, D.: Trio: enabling sustainable and scalable outdoor wireless sensor network deployments. In: IPSN 2006: Proceedings of the 5th international conference on Information processing in sensor networks, pp. 407–415. ACM Press, New York (2006)
5. El-Hoiydi, A.: Spatial tdma and csma with preamble sampling for low power ad hoc wireless sensor networks. In: Proceedings of Seventh International Symposium on Computers and Communications, 2002. ISCC 2002, pp. 685–692 (2002)
6. El Zarki, M., Liu, H., Ma, H., Gupta, S.: Error control schemes for networks: An overview. Mobile Networks and Applications 2, 167–182 (1997)
7. Hill, J., Horton, M., Kling, R., Krishnamurthy, L.: The platforms enabling wireless sensor networks. Commun. ACM 47(6), 41–46 (2004)
8. Krishnamurthy, L., Adler, R., Buonadonna, P., Chhabra, J., Flanigan, M., Kushal-nagar, N., Nachman, L., Yarvis, M.: Design and deployment of industrial sensor networks: experiences from a semiconductor plant and the north sea. In: SenSys 2005: Proceedings of the 3rd international conference on Embedded networked sensor systems, pp. 64–75. ACM Press, New York (2005)
9. Lim, R.: Wireless fire sensor network demonstrator. Master's thesis, ETH Zurich (2006)
10. Lin, E.-Y.A., Rabaey, J.M., Wolisz, A.: Power-efficient rendez-vous schemes for dense wireless sensor networks. In: 2004 IEEE International Conference on Communications, vol. 7, pp. 3769–3776 (June 2004)
11. Mainwaring, A., Culler, D., Robert Szewczyk, J.P., Anderson, J.: Wireless sensor networks for habitat monitoring. In: WSNA 2002: Proceedings of the 1st ACM international workshop on Wireless sensor networks and applications, pp. 88–97. ACM Press, New York (2002)
12. Moteiv Corporation1. Tmote Sky Ultra low power IEEE 802.15.4 compliant wireless sensor module. Datasheet (2005) (November 2006),
http://www.sentilla.com/pdf/eol/tmote-sky-datasheet.pdf
13. Piotrowski, K., Langendoerfer, P., Peter, S.: How public key cryptography influences wireless sensor node lifetime. In: SASN 2006: Proceedings of the fourth ACM workshop on Security of ad hoc and sensor networks, pp. 169–176. ACM Press, New York (2006)

14. Proxim wireless. ORiNOCO 11b Client PC Card datasheet,
    http://www.proxim.com/learn/library/datasheets/11bpccard.pdf
15. Sanyo. Twicell HR-3UTG datasheet, http://www.eneloop.info
16. Schurgers, C., Tsiatsis, V., Ganeriwal, S., Srivastava, M.: Optimizing sensor networks in the energy-latency-density design space. IEEE Transactions on Mobile Computing 1(1), 70–80 (2002)
17. Ye, W., Heidemann, J., Estrin, D.: An energy-efficient mac protocol for wireless sensor networks. In: INFOCOMM (2002)

Routing and Aggregation Strategies for Contour Map Applications in Sensor Networks

Shoudong Zou, Ioanis Nikolaidis, and Janelle Harms

Computing Science Department
University of Alberta
Edmonton, Canada
{szou,yannis,harms}@cs.ualberta.ca

Abstract. We consider the limitations inherent in data aggregation techniques for wireless sensor networks. In particular, we illustrate the limitations of schemes in which data from a sensor are routed to the node that can perform the most effective data aggregation. We then seek alternative forms of aggregation where, in principle, it is possible to aggregate readings from many sensors into a single one. To this end, we select contour maps as the most convenient example of data representation because they are used by various applications and, at the same time, allow for aggressive data aggregation. We describe particular aggregation and routing strategies that exploit the structure of contour maps. Namely, we use clustering based on the contour map values, as well as intra–cluster and inter–cluster routing and aggregation heuristics. Simulation results demonstrate the capabilities and limitations of the various varieties of aggregation/routing strategies.

1 Introduction

It is well known that in wireless sensor networks employing multi-hop forwarding, the energy of nodes closer to the sink is depleted first [1]. Solutions that cluster nodes into "neighborhoods" and perform intra–cluster aggregation do not fundamentally change this observation (think of the clusters as becoming "nodes"). Similarly, performing pairwise aggregation between two neighboring nodes is also not changing the nature of the observation (think of each pair becoming the equivalent of a single "node"). An option is to consider performing aggregation not locally, i.e., within a "neighborhood", but globally, at a node, possibly distant, someplace in the network, assuming the most beneficial site for performing aggregation is at that distant node. Such an approach trades additional load to forward to the distant aggregation site, with a potentially better reduction of the traffic that eventually marches from the aggregation site to the sink.

Assume all sensor nodes have the same transmission range, R_t, and are uniformly distributed within a 2-D area with a unique sink located at the origin. Without loss of generality, we assume the area of interest is a sector of angle $\pi/2$ (Figure 1(a)). We divide the area into k sections $A_1, A_2, ..A_k$ where the radii of each section A_i is $i * R_t$ ($i = 1, 2, ..., k$). The area of each section A_i ($i = 1, ..., k$) is $A_i = \frac{\pi}{4}(2i - 1)R_t^2 = (2i - 1)A_1$. For convenience

H. van den Berg et al. (Eds.): WWIC 2009, LNCS 5546, pp. 36–47, 2009.
© Springer-Verlag Berlin Heidelberg 2009

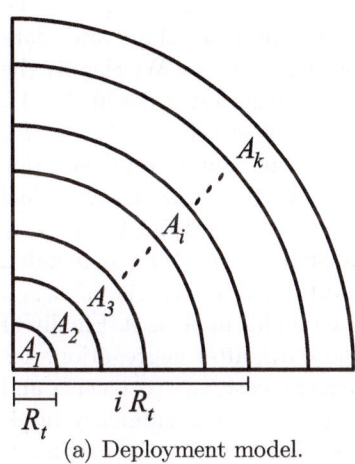

Type	Source in	Aggr. in	Energy Cost	Prob.
I	$A_2, ..., A_{i-1}$	$A_1 \cup I$	0	$\frac{\sum_{j=1}^{i-1} A_j}{A_T}$
		II	$\beta + m$	$\frac{A_i}{A_T}$
		III	$(1+m)(1+\beta)$	$\frac{\sum_{j=i+1}^{k} A_j}{A_T}$
II	A_i	$A_1 \cup I$	0	$\frac{\sum_{j=1}^{i-1} A_j}{A_T}$
		II	$\beta + m$	$\frac{A_i}{A_T}$
		III	$m(1+\beta)$	$\frac{\sum_{j=i+1}^{k} A_j}{A_T}$
III	$A_{i+1}, ..., A_k$	$A_1 \cup I$	$1+\beta$	$\frac{\sum_{j=1}^{i-1} A_j}{A_T}$
		II	$\beta + m$	$\frac{A_i}{A_T}$
		III	$m(1+\beta)$	$\frac{\sum_{j=i+1}^{k} A_j}{A_T}$

(a) Deployment model.   (b) Relay load vs. aggregation site.

Fig. 1. Model for the study of pairwise aggregation under uniform deployment

we set A_1 to be the area unit (A_1=1). The total area of the k sections is $A_T = \sum_{j=1}^{k} A_j = A_1(1 + 3 + 5 + ... + 2k - 1) = k^2$. Assume now that for each data unit generated at a sensor, there is a *best* aggregation node/site for it and the location of the best aggregation site is uniformly distributed across the whole network. We model the impact of the best aggregation by a factor, m ($0 < m < 1$), which expresses traffic reduction caused by aggregation. Specifically, for every unit of traffic (before aggregation) from two sources, we produce $1 + m$ units as a result of aggregation instead of 2. We assume, optimistically, that this "best" m applies to the aggregation of traffic from any sensor (realistically, m is not the same for all nodes). In addition, to capture the energy difference between transmitting and receiving, we define β as the ratio of the amount of energy consumption for receiving one bit over that of transmitting a bit. We assume that, typically, $0 < \beta \leq 1$.

Nodes within A_1 do not gain from aggregation because a single transmission delivers their data to the sink. Nodes in A_i ($i > 1$) impose *energy cost* on nodes in A_1, which depends on where the traffic from the node in A_i gets aggregated. For example, for a node in A_2, if its best aggregation node is in A_1, then the energy consumption it adds to nodes of A_1 is $\beta + m$ (the cost to receive it, β, and to transmit, m, the resulting aggregation outcome). If its best aggregation site is in A_j ($j = 2, 3, ..., k$) then the cost to nodes in A_1 is $(1 + \beta)m$ (since they receive and relay the already aggregated traffic). Assuming uniform distribution for the location of the best aggregation site, the expected relay load for a node in A_1 is:

$$L_1 = \left(\frac{A_T - A_1}{A_T} m(1+\beta) + \frac{A_1}{A_T} (\beta + m) \right) \frac{(A_T - A_1)}{A_1}$$

$$= ((1+\beta)mk^2 + \beta(1-m))(1 - \frac{1}{k^2}) \tag{1}$$

Next we generalize the definition of the energy cost to nodes in A_i $(i > 1)$. Besides transmitting their own sensed data[1] they also need to relay "raw" data (data before aggregation) and aggregated data for other sensors. We classify the other nodes with respect to a node in A_i into into 3 categories: nodes in A_1, A_2, ... A_{i-1}; nodes in A_i and in A_{i+1}, ... A_k, denoted respectively by $A_1 \cup I$, II, and III. Within each category the best aggregation site could be in the same or in a different category. If the aggregation site is in A_i, we assume the ideal (optimistic) case that a single transmission delivers the data within A_i and costs the aggregating node β to receive it and m to transmit the resulting aggregated traffic, for a total of $\beta + m$. If the aggregation is performed in $(A_1, A_2, ... A_{i-1})$, then no node in A_i serves as relay, and the load to nodes in A_i is 0. Finally, if the aggregation is performed in $(A_{i+1}, ... A_k)$ the data, after aggregation, has to travel to the sink and hence it imposes an energy cost $m\beta$ to receive and m to transmit on a node in A_i, for a total of $m(1 + \beta)$. The summary of all the load seen by nodes in A_i for the various categories to which the originator and the aggregator node reside is shown in Figure 1(b). Invoking the uniform distribution of the nodes in the 2-D space we can produce the weighted sum of costs, and we can determine the expected energy cost for a node in A_i:

$$
\begin{aligned}
L_i = {} & \left(\frac{A_i}{A_T}(\beta + m) + \frac{\sum_{j=i+1}^{k} A_j}{A_T}(1 + m)(1 + \beta) \right) \frac{\sum_{j=2}^{i-1} A_j}{A_i} \\
& + \left(\frac{A_i}{A_T}(\beta + m) + \frac{\sum_{j=i+1}^{k} A_j}{A_T} m(1 + \beta) \right) \\
& + \left(\frac{\sum_{j=1}^{i-1} A_j}{A_T}(1 + \beta) + \frac{A_i}{A_T}(\beta + m) \right. \\
& \left. + \frac{\sum_{j=i+1}^{k} A_j}{A_T} m(1 + \beta) \right) \frac{\sum_{j=i+1}^{k} A_j}{A_i}
\end{aligned}
\tag{2}
$$

Clearly, the closer to the sink, the larger the energy cost. To achieve maximum system lifetime nodes close to the sink should not deplete their energy any faster than any other nodes that rely on them for relaying their traffic, and since only nodes in A_2 can directly feed the nodes in A_1, the following must hold: $L_1 = L_2$. Solving $L_1 = L_2$ for m, we get:

$$
m = \frac{(1 + \beta)k^2 - (4 + 4\beta)}{(2 + 4\beta)k^4 - (1 + 9\beta)k^2 - (1 + \beta)} \qquad (k \geq 2)
\tag{3}
$$

Hence, pairwise aggregation, performed at the best possible site that would result in the most reduction of data, is *not scalable*, because for increasing k (i.e., network scale expressed as multiples of R_t) and assuming uniform sensor deployment density, the cost is justified only for an extremely aggressive data aggregation scheme (tiny m), and m must be decreasing at least as fast as k^{-2}.

---

[1] We will not account for the load of data sourced/sensed at a node, because it is an inescapable cost and it is equal to all sensors throughout the network.

2   Contour Map Aggregation

Given the limitations of pairwise aggregation, we opt for an aggregation approach which allows, in principle, aggressive n-way ($n \geq 3$) aggregation. We also consider how clusters could be "stringed" to facilitate further aggregation, i.e., without restricting the application of aggregation to within individual "neighborhoods". To do so we define aggregation applicable to contour maps. We assume the reader is already familiar with the data representation form of a contour map, which is composed of "contour lines" (also called "isolines"), an example of which is shown in Figure 2. The points and, by extension, the sensors residing between two successive contour lines are said to belong to the same level set. Contour maps are a widely utilized data representation method [2,3,4,5,6]. Between two successive contour lines, the attribute values are considered to be the same or similar, hence a contour map introduces an *approximation* which is bounded by the step size between successive contour lines.

Contour map construction from sensor data has been already addressed in [3,7], but whereas the interest in previous studies was toward discerning contour lines in the presence of noise, we are interested in the interaction between routing and aggregation for contour map data with the intent to determining the most energy efficient strategy. In summary, what we try to exploit is that given a step value (or "tolerance") that separates contour lines, sensors of the

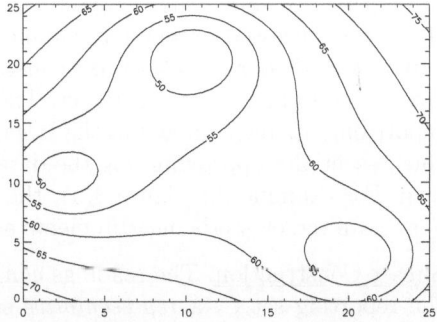

Fig. 2. Contour map example

same level set could report to the sink that they possess the exact same measurement. We assume that the position of the sensors is known, and hence, based on their positions and the attribute value they represent, the sink can reconstruct the complete contour map.

2.1   Network Architecture

We use a hierarchical architecture for aggregation purposes. Nodes are grouped into clusters, each with a cluster-head (CH) placed in charge of aggregating data of all its members (plus itself) and with sending the aggregated data to the sink, relaying the aggregated data through other CHs. As we will see, after intra–cluster aggregation, there is potential for further, inter–cluster, aggregation to be performed as well. Because of the burden to CHs, we assume a similar approach as in [8], thus "rotating" the role of CH, to even out the energy depletion within clusters.

Once a node becomes a CH, it broadcasts an advertisement message announcing its status of CH as well as its current sensor reading. Based on a given contour step-value (which we assume is globally known) and the received CH advertisement messages, each non-CH node selects as its cluster-head a CH whose sensor reading is within the same range ($[n * step, (n + 1) * step]$) of its own reading. That is clusters are formed from nodes that belong to the *same level set* and which are, of course, within communication range of each other. If there are two or more such CHs within its transmission range, the node selects the CH which is closer to it in terms of Euclidean distance. A node may fail to join any cluster because its sensor reading is not within the same range of any neighboring CH. In this case, it will elect itself the CH of a single-node cluster. After cluster construction, each cluster member reports its reading to its CH.

2.2   Intra–Cluster Aggregation

Intra–cluster aggregation is performed in two steps:

Reading Suppression: We assume sensor locations are known. Each data unit is composed of the sensor that produced it and the sensed attribute value. The objective is to collect the readings from nodes belonging to the same level set, i.e., within the same value range ($[n * step, (n + 1) * step]$). These readings can then be suppressed except for sending out a representative one. To do so, we exploit the cluster structure. By construction, all the sensor readings within a cluster are from the same level set. The aggregation performed by the CH is to report only its own value and the IDs of all its cluster members (plus its own), thus essentially approximating the data sensed at the cluster members with its own. For example, in Figure 3(a), the CH, d, which has sensed value 58, will aggregate the cluster's measurements as $< 58, C_d >$ where $C_d = \{a, c, d, e\}$.

Subset Construction: The readings can be further aggregated by choosing from the reporting set C_i a *representative* subset $C_i' \in C_i$. The selection of C_i' can be performed by making sure that all nodes in C_i are at most ϵ distance away from the set of nodes in C_i'. This is the Dominating Set (DS) [9] problem on the graph built with the vertices of C_i and with edges between any two vertices

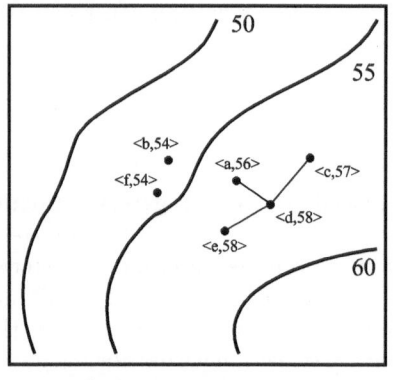

(a) Reading suppression.

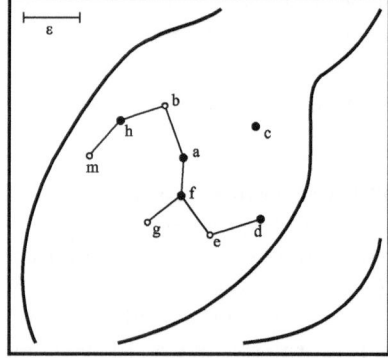

(b) Subset construction.

Fig. 3. The steps of intra–cluster aggregation

whose distance are ϵ or less. That is, we seek the DS of the graph $G(C_i, E(\epsilon))$ where $E(\epsilon) = \{(x, y) | d_{x,y} \leq \epsilon, x \neq y, x, y \in C_i\}$ where $d_{x,y}$ denotes the Euclidean distance between node x and y. Because the minimum cardinality DS is NP–Hard we employ a greedy approximation [9] which can be executed easily and locally, at the CH. The final outcome of intra–cluster aggregation is $< s_i, C_i' \cup \{i\} >$, which includes the CH i (if not already in the DS), as it is the CH's value s_i that represents all the sensors in the cluster. Figure 3(b) provides an example where all nine nodes are within range of each other and form a cluster with CH the node a. The intra–cluster aggregation produces $< s_a, \{a, c, d, f, h\} >$. Note that edges in Figure 3(b) represent the set $E(\epsilon)$.

Note that, based on the $C_i' \cup \{i\}$ received from each CH, and because the sink is aware of ϵ and the locations of all nodes, the sink can determine which nodes are dominated by each C_i' assuming that nodes that belong to different level sets are at least ϵ units away. If they are any closer, then the subset construction essentially introduces a spatial approximation on top of the value approximation that resulted from the reading suppression step. Moreover, if two CH, say i and g, formed by nodes that are on the *same* level set, could coordinate their subset construction, they could do so by determining the DS of their *joint* graph, i.e., the union of their two ϵ–based graphs, plus edges of distance ϵ or less between vertices of the two clusters. Naturally, the construction of the DS of the combined graph makes sense when the clusters are close to each other, i.e., they have at least two nodes at a distance ϵ or less. We exploit this possibility in the second part, that of inter–cluster aggregation.

2.3   Inter–Cluster Aggregation

In inter–cluster aggregation the task is to determine pairs of clusters, such that their combined aggregated data can be aggregated further. In order for this to happen, the two clusters must be formed by nodes that observed measurements that fall within the same level set. In addition, if the clusters are close to each other, then the subset construction outlined in the previous paragraph can also be applied, but its impact is expected to be, in general, rather limited. Hence, for every two clusters, one can define the *potential* of the aggregation outcome numerically. Namely, in order to evaluate the amount of data reduction between a pair of CHs, we define a *joint flow loss multiplier* (m_{xy}) between CHs x and y, as the data reduction possible if the two node sets of the two CHs are combined and collectively represented by the dominating set of the graph defined by their union. (For well–separated clusters, this will be the union of their DS.) Thus, if the cluster with CH i sends its aggregated data $< s_i, C_i >$ to CH g which possesses the aggregated data $< s_g, C_g >$, and if D_i represents the "size" of the aggregated data from i, i.e., $D_i = | < s_i, C_i > |$ then the post–aggregation data at g will be of size $D'g = (D_g + D_i) * m_{ig}$.

At this point, the problem becomes an optimization problem: which clusters to pair so that their aggregate traffic is the least possible. Furthermore, after aggregating the traffic of two clusters, we can split it to multiple flows and forward it to the sink, thus attaining a degree of load balance. Essentially, the problem becomes one that we have already discussed in [10] where first the aggregation is determined and then the resulting aggregated data are split using a flow model. To reduce the computational complexity for solving the flow problem and make our approach scalable, we consider CHs as nodes representing the collected aggregated data. The particular flow fragments and the paths taken to the sink can be computed via our IP/LP model outlined in [10]. However, we first need to determine the pairs of clusters, and to do so we employ a variety of heuristics:

Maximum Aggregation Perfect Matching (MAPM): MAPM attempts to maximize the total inter–cluster data reduction across the entire network. The joint flow loss multiplier is utilized to measure the benefit from inter-cluster aggregation over each pair of CHs. We use $(D_i + D_g) * (1 - m_{ig})$ to denote the total joint volume reduction via aggregation and place it as edge weight on a graph including as vertices only the CHs. We apply maximum aggregation perfect matching [11] across the network. After perfect matching and inter-cluster aggregation, each pair of matched CHs (i and g) is treated as a single data flow whose data demand is $(D_i + D_g) * m_{ig}$ generated at g. If the number of clusters is odd then we are left with single CH that does not employ inter–cluster aggregation.

Local Maximum Aggregation Perfect Matching (Local MAPM): This is a variation of MAPM where pairs have to be neighbors, on the assumption that traveling a longer path to a better aggregation opportunity is expensive in terms of energy.

ENergy Critical node Aware Spanning Tree (ENCAST): ENCAST is introduced in detail in [12], and contrary to the previous two schemes it does not split aggregated data into flows. Instead it builds a single tree. The tree is reconstructed, i.e., a new "version" of the tree is built, as nodes become critically low in terms of energy. Critical nodes become leaf nodes of the tree. A parameter, α, is used to trigger tree reconstructions depending on the energy depletion situation in the network. Specifically reconstruction occurs when the residual energy of any node drops below the threshold given by $\alpha^k E$ where E is the initial energy and $k = 1, 2, ...$ indicates the "version" of the tree. Thus the longer the network operates (and the more times the tree is reconstructed) the lower the threshold is pushed. Because the routing to the sink is restricted to the tree, the inter–cluster aggregation is not performed as in the previous two schemes, but progressively as the aggregated data travel to the sink from one CH to the next, along the tree. When the aggregated data arrive at a CH of a cluster from the same level set, they are aggregated and sent out. This form of aggregation can result in more than two CHs aggregating their data together.

Finally, we also consider a benchmark scheme (we will call it General) which represents intra–cluster aggregation *alone* without any inter–cluster aggregation performed. Subsequent to intra–cluster aggregation, the data are split into multiple flows and routed to the sink as per the solution of an LP flow optimization.

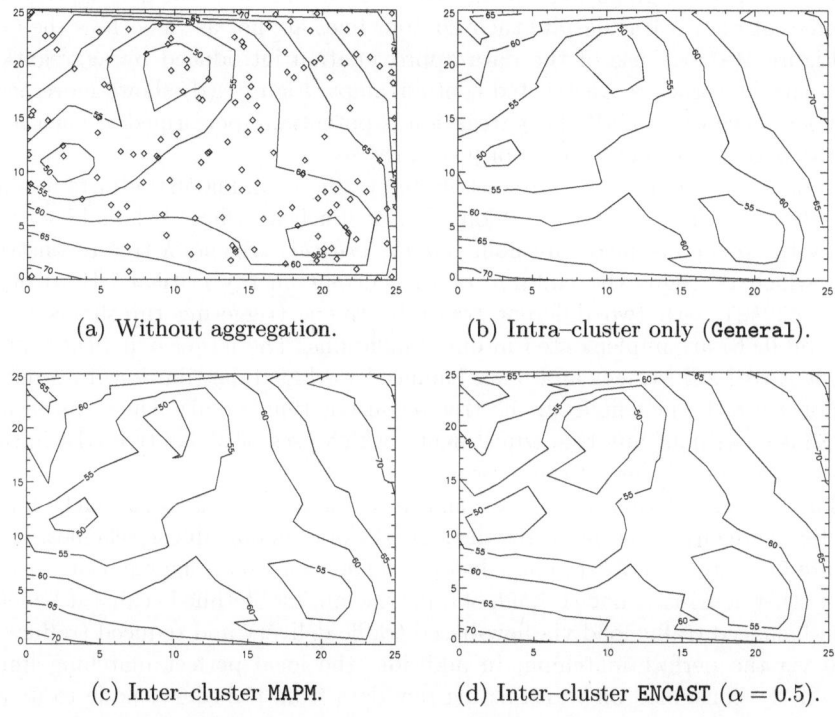

(a) Without aggregation.   (b) Intra–cluster only (**General**).

(c) Inter–cluster **MAPM**.   (d) Inter–cluster **ENCAST** ($\alpha = 0.5$).

Fig. 4. Quality of reconstruction for various aggregation options

3   Performance Evaluation

We simulate a sensor network with anywhere from 60 to 140 nodes placed in
a $25 \times 25 m^2$. The location of the 140 nodes is shown as dots in Figure 4(a).
The sink is placed at the origin (0,0). Nodes start with energy of 10 Joules and
each of them generates sensor data at a constant rate of 1 Kb/sec. We adopt
the energy consumption model of [13]. Energy spent on receiving and sending
data packet is determined by the wireless communication environment and the
packet size. We assume each sensor node's wireless transmission range is 10 m.
The contour data are dynamically generated by the diffusion model presented in
[14] which ensures the slow but continuous change of the contours. To show the
quality of the contour data collected (using 140 nodes) we present reconstructed
contour maps (Figure 4) the way the sink would generated them. Namely, at
the sink, the per-sensor data are reconstructed from the received aggregated
data, interpolated, smoothed, and presented as a contour map using a data
visualization tool [15]. The step-value is 5 and $\epsilon = 2m$.

Comparing Figure 4(a) to the benchmark of Figure 2, the degradation mainly
comes from less deployed sensors and random sensor placement (meaning poor
sensing coverage and coarse granularity). However, from Figure 4(b) and 4(c), we

observe that our aggregation schemes are visually similar to the non-aggregation schemes of Figure 4(a) despite their contour lines are not as smooth as the ones in Figure 4(a) because of the data approximation introduced by aggregation. Compared to other reconstructed contour maps, Figure 4(d) shows more accuracy loss because in ENCAST, aggregation is potentially performed at each node on the path to the sink instead of just pairwise.

In order to balance unevenly distributed traffic load, ENCAST adjusts its data collection tree in a proactive fashion. The cluster head whose residual energy is below the predetermined threshold (*energy critical*) triggers a tree reconstruction which reassigns the children of the current energy critical CH to other CHs. ENCAST with two different tree adjustment triggering thresholds ($\alpha =$ 0.5 *and* 0.66) are implemented in our simulations. The larger α indicates more frequent tree adjustment and more balanced workload distribution. In order to demonstrate the benefit from the tree reconstruction, we also show the results for ENCAST without any tree adjustments, which uses only one tree (the initial) throughout the lifetime of the network.

The network lifetime results are shown in Figure 5(a). General lacks inter–cluster aggregation and performs the second worst among all the schemes. From Figure 5(a), we also observe flow-based schemes with data aggregation outperform other heuristics. Local MAPM obtains the longest lifetime because it benefits from balanced traffic load via flow-based traffic delivery and reduced traffic volume via the perfect matching. In addition, the local perfect matching limits the energy consumed on transmitting raw data from the source node to its aggregation site which may be far away from the source. This is also the reason the lifetime achieved by Local MAPM is longer than MAPM. MAPM performs perfect matching without any constraints. It reduces larger traffic volume than Local MAPM with more energy consumed along potentially longer paths.

ENCAST with various α's attains lifetime similar to the flow-based schemes despite not splitting the traffic flow. Tree reconstruction removes the heavy burden of data routing from energy critical CHs, essentially balancing energy consumption among all the CHs. Due to data aggregation, ENCAST ($\alpha = 0.66$) and ENCAST ($\alpha = 0.5$) perform better than General in terms of lifetime. Among all the schemes, ENCAST without tree adjustment achieves the minimum lifetime. This confirms that ENCAST with tree adjustment is a very effective technique to extend system lifetime. We also show the total data volume collected by each heuristic in Figure 5(c). Aggressive on-tree aggregation, ENCAST ($\alpha = 0.66$) and ENCAST ($\alpha = 0.5$) reduces the volume of data delivered to the sink compared to flow–based approaches.

Since ENCAST starts with a hop-based shortest path tree the tree reconstruction tends to increase the depth of the tree and therefore data traveling from its CH to the sink have more chances to be aggregated at another CH along the way. The collected data volume decreases (worsens) as α increases. MAPM and Local MAPM perform perfect matching to pair CHs for aggregation and gather similar amount of data to ENCAST's with tree adjustment ($\alpha > 0$). Naturally, the total amount of collected data increases as the sensor density increases.

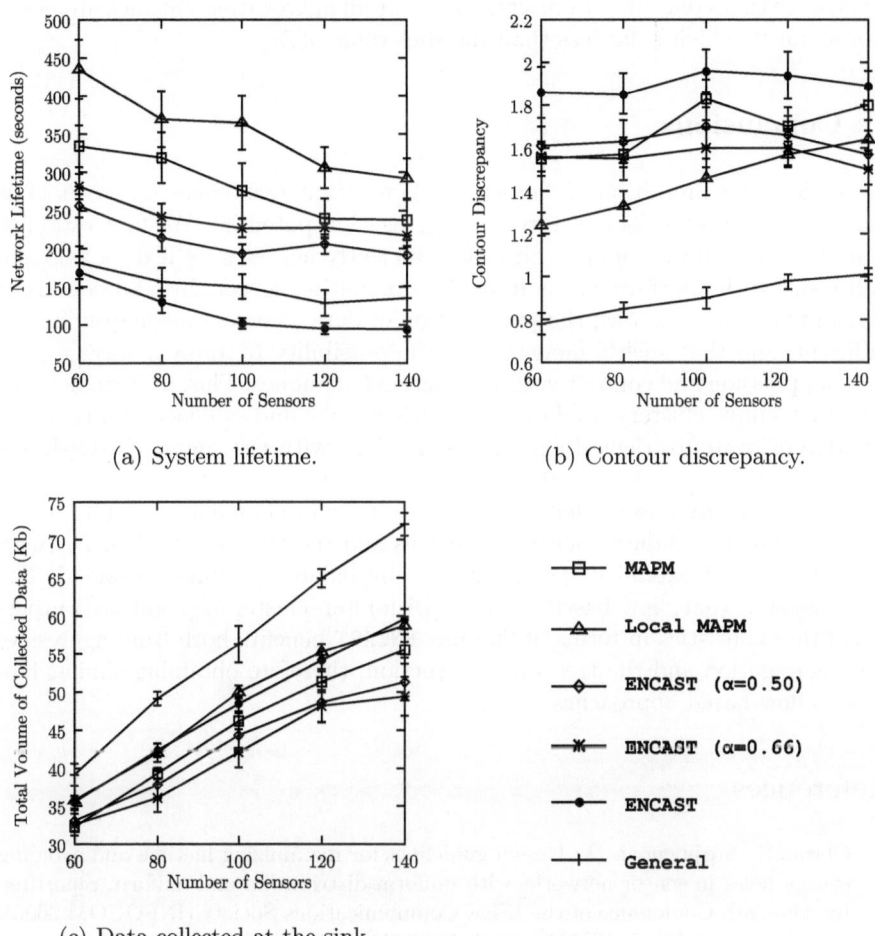

(a) System lifetime.

(b) Contour discrepancy.

(c) Data collected at the sink.

Fig. 5. Simulation results

To evaluate the collected data accuracy, we introduce another metric similar to [16]: *contour discrepancy*. We examine the reconstructed (at the sink) data, s'_i, for each sensor, i, compared to the actual data, s_i, and formulate the average contour discrepancy as: $(\sum_{i=1}^{N} |s'_i - s_i|)/N$. Data from certain sensors that were eliminated during the aggregation are replaced by the data values from sensors that dominate them. If the missing sensor's value is dominated by more than one node sensor, then the value is chosen randomly from one of the dominators.

We present contour discrepancy for different aggregation strategies in Figure 5(b) where the units on the Y-axis are the same as those of the step-value. General does not apply inter–cluster aggregation, so it maintains the largest data accuracy. MAPM performs a more aggressive aggregation than Local MAPM and results in more accuracy loss. For tree–based schemes, they all perform aggressive data aggregation and have large contour discrepancy. However, we observe that contour data discrepancies for all aggregation schemes are below 2 or about 2, which is far less than the step-value of 5.

4   Conclusions

We review the limitations inherent in aggregation techniques where data is routed to the best sensor in terms of aggregation potential. We first establish that there are limited opportunities for efficiently aggregating if data aggregation needs to be performed far from the originating sensor. Instead, we argue that in many applications, the presentation of data as a contour map might be sufficient, and that such a model opens the possibility to apply n-way ($n \geq 3$) data suppression and contour vector reduction techniques. This, in turn, helps us construct simple cluster-based routing heuristics. We find evidence that the combination of contour-oriented aggregation models, with the simple cluster-based routing heuristics, leads to scalable data collection.

There are many lessons that can be drawn from our simulation results. First, both the flow-based data collecting schemes and the tree-based ENCAST's (with tree adjustment) extended system lifetime by balancing traffic workload. The other lesson is that, flow-based schemes (with inter–cluster aggregation) outperform other heuristics in terms of lifetime. ENCAST benefits both from aggressive data aggregation and the tree's reconstruction, therefore obtaining similar lifetime to flow-based approaches.

References

1. Olariu, S., Stojmenovic, I.: Design guidelines for maximizing lifetime and avoiding energy holes in sensor networks with uniform distribution and uniform reporting. In: The 25th Conference of the IEEE Communications Society (INFOCOM 2006), Barcelona, Catalunya, SPAIN, pp. 1–12 (2006)
2. Solis, I., Obraczka, K.: Efficient continuous mapping in sensor networks using isolines. In: The 2nd Annual International Conference on Mobile and Ubiquitous Systems: Networking and Services (MobiQuitous 2005), San Diego, California, USA, pp. 325–332 (2005)
3. Liao, P.K., Chang, M.K., Kuo, C.C.: Contour line extraction with wireless sensor networks. In: The 2005 IEEE International Conference on the Communications (ICC 2005), Seoul, Korea, pp. 3202–3206 (2005)
4. Buragohain, C., Gandhi, S., Hershberger, J., Suri, S.: Contour approximation in sensor networks. In: Gibbons, P.B., Abdelzaher, T., Aspnes, J., Rao, R. (eds.) DCOSS 2006. LNCS, vol. 4026, pp. 356–371. Springer, Heidelberg (2006)
5. Xue, W., Luo, Q., Chen, L., Liu, Y.: Contour map matching for event detection in sensor networks. In: The 2006 ACM SIGMOD International Conference on Management of Data, Chicago, Illinois, USA, pp. 145–156 (2006)
6. Gandhi, S., Hershberger, J., Suri, S.: Approximate isocontours and spatial summaries in sensor networks. In: The 6th International Conference on Information Processing in Sensor Networks (IPSN 2007), Cambridge, Massachusetts, USA, pp. 400–409 (2007)

7. Liao, P.K., Chang, M.K., Kuo, C.C.: A distributed approach to contour line extraction using sensor networks. In: The 2005 IEEE Vehicular Technology Conference (VTC 2005), Dallas, Texas, USA, pp. 2716–2720 (2005)
8. Heinzelman, W.R., Chandrakasan, A., Balakrishnan, H.: Energy-efficient communication protocol for wireless microsensor networks. In: The 33rd Hawaii International Conference on System Science (HICSS 2000), Maui, Hawaii, USA, pp. 3005–3014 (2000)
9. Hochbaum, D.S.: Approximation Algorithms for NP-Hard Problems. PWS Publishing Company, Boston (1997)
10. Zou, S., Nikolaidis, I., Harms, J.: Efficient aggregation using first hop selection in wsns. International Journal of Sensor Networks 4(13), 55–67 (2008)
11. Liu, H., Jia, X., Wan, P., Yi, C.W., Makki, S.K., Pissinou, N.: Maximizing lifetime of sensor surveillance systems. IEEE Transactions on Networking 15, 334–345 (2007)
12. Zou, S., Nikolaidis, I., Harms, J.: ENCAST:energy–critical node aware spanning tree for sensor networks. In: The 3rd Annual Conference on Communication Networks and Services Research (CNSR 2005), Halifax, Nova Scotia, Canada, pp. 249–254 (2005)
13. Feeney, L.M., Nilsson, M.: Investigating the energy consumption of a wireless network interface in an ad hoc networking environment. In: 20th Conference of the IEEE Communications Society (INFOCOM 2001), Anchorage, Alaska, USA, pp. 1548–1557 (2001)
14. Meng, X., Nandagopal, T., Li, L., Lu, S.: Contour maps: Monitoring and diagnosis in sensor networks. Computer Networks 50, 2820–2838 (2006)
15. ITT Visual Information Solutions Image Processing and Data Analysis: IDL 7.0, http://www.ittvis.com/ProductServices/IDL.aspx
16. Solis, I., Obraczka, K.: Isolines: efficient spatio-temporal data aggregation in sensor networks. Wireless Communications and Mobile Computing 9, 357–367 (2007)

Path-Based Reputation System for MANET Routing

Ji Li, Teng-Sheng Moh, and Melody Moh[*]

Department of Computer Science
San Jose State University, San Jose, CA 95192-0249, USA
moh@cs.sjsu.edu

Abstract. Most existing reputation systems in mobile ad hoc networks (MANET) consider only node reputations when selecting routes. Therefore, reputation and trust are only ensured within a one-hop distance when routing decisions are made. This often fails to provide the most reliable, trusted route. In this paper we propose a system that is based on *path reputation*, which is computed from the *reputation* and *trust* values of each and every node in the route. This greatly enhances the reliability of the resulting routes. The system is simulated on top of the AODV (Ad-hoc On-demand Distance Vector) routing protocol. It is effective at detecting misbehaving nodes, including selfishness and worm-hole attacks. It greatly improves network throughput in the midst of malicious nodes and requires very limited message overhead. To our knowledge, this is the first *path-based* reputation system proposal that is applicable to non-source-based routing schemes.

Keywords: AODV, mobile networks, reputation, routing, trust.

1 Introduction

Mobile ad hoc networks (MANET) are communication networks in which nodes can dynamically establish and maintain connectivity with each other. Each node can also act as a router to forward packets on behalf of other nodes. Their major advantages include low cost, simple network maintenance, and convenient service coverage. These benefits, however, come with a cost.

Due to the lack of control over the other nodes in the network, selfishness and other misbehaviors are possible and easy. Therefore, a major challenge is ensuring security and reliability in these dynamic and versatile networks. One approach uses a public key infrastructure to prevent access to nodes that are not trusted, but this central authority approach reduces the ad-hoc nature of the network.

Reputation systems in MANET are intended to address this issue. They are effective at detecting insider attacks [12], such as selfishness in packet dropping, worm-hole attacks, and forge replies. Most existing proposals, however, consider only neighbors' reputations when selecting routes. In such systems, each node selects the next hop from its neighbors based on their reputation and trust values. Thus, trustworthiness is only ensured in a one-hop distance when making routing decisions.

---

[*] Corresponding author.

H. van den Berg et al. (Eds.): WWIC 2009, LNCS 5546, pp. 48–60, 2009.

These greedy approaches [1, 2, 3, 4, 6, 10, 14, 15] usually do not provide the most reliable routes, as the decisions are based on local information.

In this paper we propose a reputation system that maintains *path* reputation based on the reputation and trust of *every node* in the path. We use innovative ways to increment and decrement trust values corresponding to positive and negative observations, respectively. We also consider a range of values (instead of absolute values) for trust to weigh second-hand information to avoid unnecessary transient fluctuations. As a result, the system is effective at detecting misbehaviors and ensuring efficient routing. This is, to the best of our knowledge, the first *path-based* reputation system proposed that is not for source-based MANET routing (such as DSR, the Dynamic Source Routing protocol). We believe that the path-based approach may be applied to other node-based reputation systems [1, 2, 3, 4, 6, 10, 14, 15] to improve their overall effectiveness.

2 Related Studies

Due to page limits, we give a brief overview on related works; details may be found in the original papers and a recent survey paper [11]. Reputation systems may be broadly classified into two groups. The first group, *one-layer reputation systems*, refers to those that are based on observations (both direct and indirect) without an explicit evaluation of indirect observations (i.e., second-hand information). Key examples include Watchdogs and Pathraters [9], CORE (COllaborative REputation) [10], OCEAN (Observation-based Cooperation Enforcement in Ad-hoc Networks) [2], SORI (Secure and Objective Reputation-based Incentive Scheme for Ad-hoc Networks) [5], and LARS (Locally Aware Reputation System) [6]. All of them were either designed for or evaluated over DSR.

The second group, *two-layer reputation systems*, refers to those based not only on observations, but also on *trust*, which evaluates the trustworthiness of second-hand information. Major examples include two that were designed for DSR and three for AODV. The first two were CONFIDANT (Cooperation Of Nodes: Fairness In Dynamic Ad-hoc NeTworks) [3, 4] and SAFE (Securing pAcket Forwarding in ad hoc nEtworks) [15]. The three designed for AODV were TAODV (Trusted AODV) [8], Cooperative and reliable AODV [1], and Trusted-based security framework for AODV [14].

None of the above systems, however, considered *path reputation*. As mentioned in the Introduction section, most of them selected the path based on the greedy approach; i.e., choosing the most reliable next-hop from its neighbor nodes. Among the three for AODV, both TAODV [8] and Trusted-based security framework for AODV [14] used the greedy approach. The Cooperative and reliable AODV [1] used their trust model to decide the need for authenticating nodes during route discovery and route maintenance.

The only existing reputation scheme that considered *path metric* was the proposal of Watchdogs and Pathraters [9], an early work on one-layer reputation systems. Each node used Watchdog to maintain a *rating* of every other node it knew about, and used Pathrater to compute a path metric by *averaging* the node ratings in the path.

Note, however, that it could only be implemented on top of source-routing protocols such as DSR, since it relied on the Pathrater to know the exact path.

Below, we provide a simple description of AODV [13]. It is an on-demand routing protocol designed for MANET. The route is initiated by broadcasting the Route Request (RREQ) message from the source node, and the message is then propagated through the entire network. When the destination node or an intermediate node that has a route receives the RREQ, it responds with a Route Reply (RREP) message. Once the source receives the RREP, it may begin to send packets to the destination.

3 Path-Based Reputation System

3.1 Main Ideas

In the following, we present the major ideas of the proposed scheme:

(A) Two Levels of Information: First-hand and Second-hand
The proposed system makes use of both first-hand and second-hand information. First-hand information comprises direct observations done by each node on its neighbor nodes on packet forwarding. Second-hand information arrives from *indirect observations*; i.e., a neighbor node's observations of other nodes. Note that RREQ and RREP (AODV control packets) are also considered second-hand information since they carry path reputation values (see Section 3.3 for details). All second-hand information will go through a *trust* test, which is discussed next.

(B) Two Levels of Credibility: Reputation and Trust
The proposed system contains two levels of credibility. *Reputation* tells whether the node behaves correctly in the base system (routing protocol) while *trust* evaluates whether the node lies in the reputation system. Specifically, the trust system assesses the credibility of second-hand information sent by a neighbor.

(C) Two Levels of Reputation: Node-Based and Path-Based Reputation
In addition to the node reputation that is formed by both first- and second-hand information, the proposed system computes path reputation based on the node reputations of *all the nodes in the path*. The routing decision is based first on path reputation. Then, if there is a tie, it is on path distance. If there is a tie again, it can be based on the ID of the next hop. Note that routes that have very low path reputation are discarded even during route discovery phase (by removing RREQ or RREP packets). This approach significantly improves overall throughput and reduces protocol message overhead.

(D) Range-Based vs. Value-Based Reputation and Trust
In computing node and path reputation values, it is important to properly consider the trust factor of second-hand information. Instead of using an absolute trust value, the proposed system makes use of *ranges*. That is, if two trust values fall within the same range, they are assigned the same level of trust (same weight). This avoids unnecessary adjustments caused by transient fluctuations that are quite frequent in MANET.

3.2   System Architecture

In this subsection, we present a high-level view of the system design. Major components will be described in detail in the next subsection. Figure 1 shows the architecture of the proposed system. It includes five components, described below. The arrows indicate interactions among these components. Note that solid arrows are specifically for exchanges of reputation system messages, including direct observations (such as suspicious events), indirect observations (also called ALARM), and trust values. Dotted arrows are for routing control messages RREQ and RREP (they also carry path reputation values). Note also that an extra field is needed on RREQ and RREP messages to hold path reputation values.

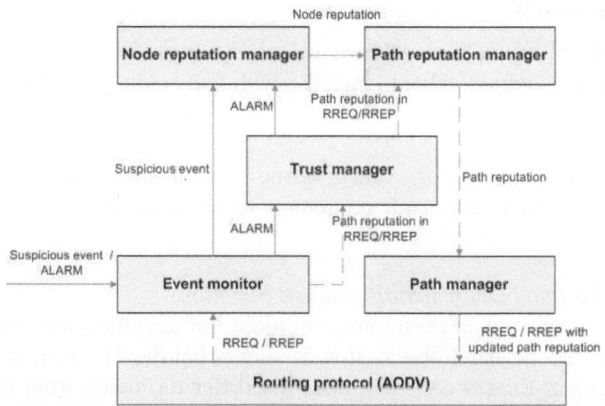

Fig. 1. System Architecture

1. *Event monitor*: It works like a watchdog [9] that detects suspicious events (such as packet drops and incorrect RREQ) and feeds them to the Node reputation manager. It also receives second-hand information (i.e., ALARM messages) from neighbors, and routing control messages by interacting with the underlying routing protocol (AODV or another routing protocol). It then feeds these indirect observations to the Trust manager.

2. *Trust manager*: It evaluates the trust (trustworthiness) of second-hand information provided by the Event monitor (see section 3.3.1). It then feeds the evaluated information and trust values to either node or path reputation managers.

3. *Node reputation manager*: It computes neighboring node reputations (see section 3.3.2). This is based both on direct observations passed by the Event monitor and on evaluated indirect observations and trust values passed by the Trust manager.

4. *Path reputation manager*: It handles path reputations based on neighbor node reputations (passed by the Node reputation manager) and second-hand path reputations and trust values (passed by the Trust manager); see section 3.3.3 for details.

5. *Path manager*: It makes routing decisions based on path reputation values fed by the Path reputation manager and passes the decision to the underlying routing protocol.

3.3 Major Components

In this section, three major components are described in detail. We first list the main variables needed to execute the proposed scheme in a network node:

- D_n: **Direct** observation of node n.
- $I_{n,m}$: **Indirect** observation of node n made by node m
- R_n: **Reputation** of node n based on direct observation
- S_n: Reputation of node n based on **Second**-hand information
- T_n: **Trust** value of node n
- PR_p: **Path Reputation** of route p
- $PS_{p,m}$: **Path** reputation of route p received from node m (considered as **Second**-hand information)

Note that all the reputation and trust values are normalized between 0 and 1; this can prevent path reputation value to increase with hop count in the route.

3.3.1 Trust Manager

Recall that it is responsible for evaluating the *trust* of second-hand observations, as well as that of second-hand path reputation reported in routing control messages (RREP).

(A) Deviation Test and Decrement/Increment Functions

First, in order to prevent false alarms, we adopt the deviation test method [3, 4] to evaluate a new second-hand observation as shown below. The idea is that a second-hand observation is trusted only if it does not differ too much from the node's own direct observation. In this case, the trust is increased. Otherwise, it is decreased. The increment and decrement are each further guided by a quadratic function to be described below.

Following is the condition to pass the deviation test:

$$|D_n - I_{n,m}| < d$$

In the above, d is the threshold for the test ($d = 0.1$ is used in this work). If the difference between D_n (the direct observation) and $I_{n,m}$ (the new indirect observation given by node m) is greater than or equal to d, then $I_{n,m}$ fails the deviation test, and T_m, the trust value of node m that sent this second-hand observation, is decreased by a decrement function, f_D (refer to Figure 2a):

$$f_D(T_m) = \frac{1}{2}(T_m)^2 \tag{1}$$

i.e., $T_m = T_m - f_D(T_m)$

On the other hand, if the difference between D_n and $I_{n,m}$ is less than d, then the node m passes the test, and T_m is increased by an increment function, f_I (refer to Figure 2b):

$$f_I(T_m) = T_m - \frac{1}{2}(T_m)^2 \tag{2}$$

that is

$$T_m = T_m + f_I(T_m)$$

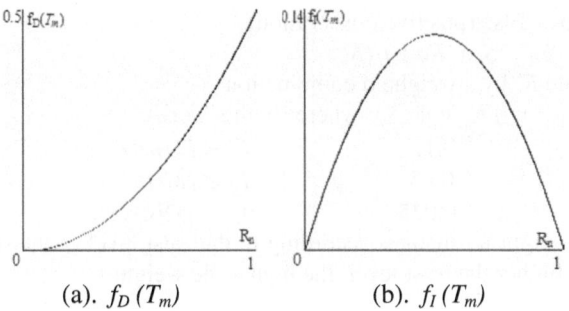

(a). $f_D(T_m)$ (b). $f_I(T_m)$

Fig. 2. Decrement and Increment Functions

The quadratic function f_D is used such that the decrement of trust is small when the trust value is still small. Yet, when the trust value is large, the corresponding decrement needs to be larger to "punish" the bad behavior more severely. The increment function f_I may be similarly explained. Note that these two functions are also used in node and path reputations.

(B) Range-Based Trust Values
Secondly, to prevent transient fluctuations, we use categorized (or range) values for trust (instead of absolute values). In our implementation, the trust value is categorized into three ranges:

$$T_m \in \begin{cases} Trust, & 1 \geq T_m > 0.7 \\ Mistrust, & 0.7 \geq T_m > 0.3 \\ Distrust, & 0.3 \geq T_m > 0 \end{cases}$$

This T_m value is then fed to either the Node or Path reputation managers to aid their evaluations of node or path reputations.

3.3.2 Node Reputation Manager

The node reputation manager computes node reputation based on both direct observations from the Event monitor and the *trust-evaluated* indirect observations from the Trust manager. For each new direct and indirect observation concerning a node n, it computes the new reputation value of node n by making use of the decrement and increment functions introduced above (in Equations 1 and 2), and by using a *weighted* combination of direct and indirect observations, as follows:

1. If the new observation is a direct observation
    a. If the observation is negative
        Then $R_n = R_n - f_D(R_n)$
    b. Else (positive observation)
        $R_n = R_n + f_I(R_n)$
2. Else (indirect observation, say $I_{n,m}$)
    If T_m is trusted **or** $I_{n,m}$ passed the deviation test at the Trust manager
    a. If the observation is negative
        Then $S_n = R_n - f_D(R_n)$

      b.   Else (positive observation)
$$S_n = R_n + f_1(R_n)$$
3.   Update R_n by a weighted combination
$$R_n = (1 - w_1) R_n + w_1 S_n, \text{ where (in this work)}$$

$$w_1 = f(T_m) = \begin{cases} 0.1, & T_m \in \{Trust\} \\ 0.05, & \text{if} \quad T_m \in \{Mistrust\} \\ 0.025, & T_m \in \{Distrust\} \end{cases}$$

Note that the weight w_1 changes according to the trust level of the second-hand information S_n; the higher the trust level, the higher the weight.

3.3.3 Path Reputation Manager

When the Path reputation manager obtains a new path reputation value (say passed by node n) from the Trust manager, it also receives T_n (the trust level of node n). In addition, it obtains R_n (reputation of node n) from the Node reputation manager. Based on these values, it computes the new path reputation value, PR_p, by taking the minimum of R_n and a *weighted* value of $PS_{p, n}$, as follows:

$$PR_p = \text{minimum} [R_n, w_2 \times PS_{p, n}] \tag{3}$$

Where the weight, w_2, is a function of T_n and is defined below in our implementation; again, note that it changes according to T_n (the trust level of node n).

$$w_2 = \begin{cases} 1, & T_n \in \{Trust\} \\ 0.5, & \text{if} \quad T_n \in \{Mistrust\} \\ 0.25, & T_n \in \{Distrust\} \end{cases}$$

If, however, the new PR_p falls below a pre-defined threshold value (Th_{path}), then the RREP or RREQ is dropped. This ensures that only trusted RREP and RREQ packets are forwarded.

3.4 Base Case: Node-Based Reputation System

In order to understand the advantages of the proposed path-based reputation system, we also implemented a node-based reputation system in the simulation, which will be used as the base case for comparison. It is identical to the proposed system except that Equation (3) is simply:

$$PR_p = R_n \tag{4}$$

Thus, each node in the network selected route is based solely on the neighbor's (the potential next hop's) reputation rather than a path-based reputation. This is similar to the greedy approach used by other schemes [1, 14].

3.5 Protocol Overhead and Complexity Analysis

In this section we note that the additional overhead required by the proposed system has the same complexity as AODV; this includes time (for computation), space (for storage), and message (for extra control packets) complexities. A detailed analysis may be found in a complete technical report [7]. Message complexity (overhead) is also evaluated by simulation in the next section.

4 Performance Evaluation

4.1 Simulation Settings

The simulation was developed on GloMoSim 2.03 [16] with the standard AODV implementation. Three protocols were compared: the original AODV, AODV with node-based approach, and AODV with path-based approach. This enabled us to observe and assess the advantages of the path-based scheme against the greedy, node-based approach.

The simulated network contained 50 nodes distributed uniformly in a 1600 meters by 1600 meters area. Each simulation experiment lasted for 10 minutes. The mobility mode was set to the Random Waypoint Model in which nodes moved to a random destination at a speed uniformly distributed between 0 and 5 m/sec and stayed at this destination for 20 sec. Misbehaving nodes were randomly selected, excluding source and destination nodes. Both UDP (supporting CBR) and TCP (supporting FTP) were evaluated, each with 10 pairs of source and destination nodes chosen randomly. Finally, $Th\_alarm = 0.49$ and $Th\_route = 0.51$ for the reputation system.

4.2 Misbehaviors and Performance Metrics

While there are many possible MANET misbehaviors, in the simulation we implemented three major MANET routing misbehaviors [12] that may be effectively dealt with by a reputation system:

- Data selfish (or selfish in data forwarding) – A misbehaving node does not forward any data packets.
- Forge reply [12] (or worm-hole attacks) – A misbehaving node replies to a *RREP* message with hop-count equal to 1 for any incoming *RREQ* message. Thus, the node claims that it has the shortest path to the destination.
- A combination of forge reply and data selfish.
- Forge data– A misbehaving node modifies data packets that pass through.

Note that a forge reply attack directly affects (or fakes) a routing path (also called route disruption [12]) since it replies with incorrect routing information. On the other hand, data selfish or forge data misbehavior affects a single node. Simulation results showed that the proposed *path-based reputation scheme is most effective when dealing with routing path attacks*.

The performance metrics include *throughput, average end-to-end packet delay,* and *reputation message ratio* - the percentage of reputation system messages in the overall control messages (including those for routing). This represents the reputation system overhead. Note that reputation message overhead is also an effective indicator of extra *energy consumption* due to transmissions.

4.3 Simulation Results: CBR over UDP

We first set the network application to CBR with 512 bytes packets continuously sent every 200 msec, resulting in approximately 20.5 kbps for each of the 10 flows, or a total of 205 kbps in the network.

(A) Misbehavior: Forge Reply

Simulation results are shown in Figure 3, including three sub-figures: throughput, delay, and reputation message ratio; each shows three schemes: reputation-disabled, node-based reputation, and path-based reputation. These results indicate that the path-based reputation system is very effective at dealing with forge reply attack. It has more than doubled the throughput as compared to node-based reputation, and more than tripled when compared with the original AODV. On the other hand, as its throughput is very high, more packets are queued at intermediate nodes, resulting in a long end-to-end delay.

Note that the reputation message overhead for the path-based reputation system has stayed very low. The main reason is that when a RREQ or RREP packet has a path reputation value that falls below *Th_route,* it is discarded immediately. This mechanism successfully prevents many control packets with a low path reputation value from traversing further down the path, and therefore greatly reduces control overhead. On the other hand, the node-based reputation system obviously does not have this advantage – it does not check the reputation value along the path and therefore has a much higher overhead. The important advantage of low reputation overhead for the path-based scheme has remained true in all the experiments (Figures 3, 4, 5, and 6).

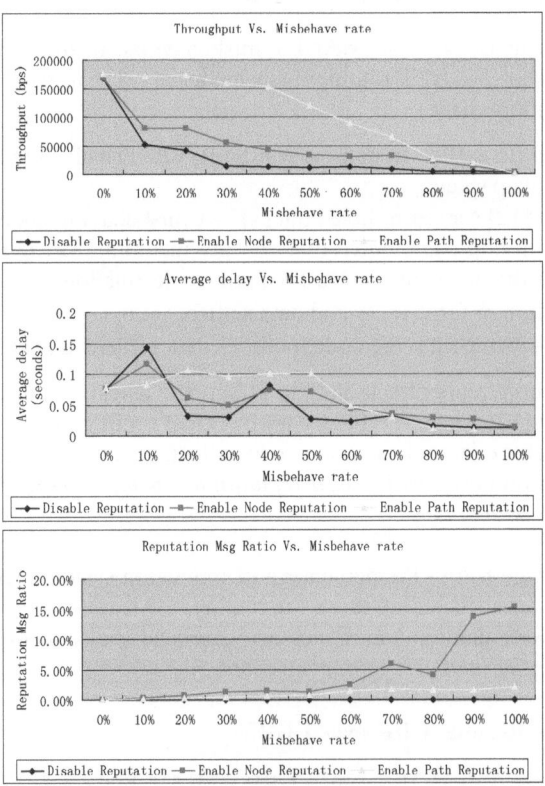

Fig. 3. Performance in Forge Reply (UDP)

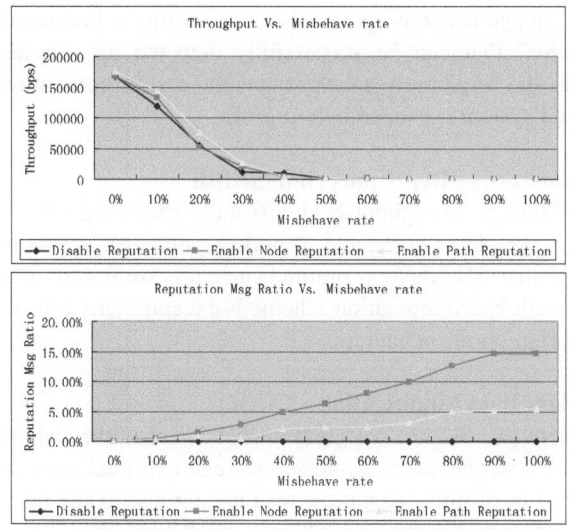

Fig. 4. Performance in Data Selfish (UDP)

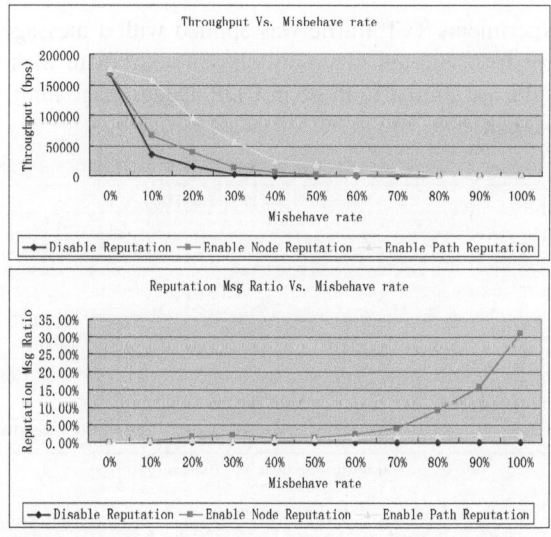

Fig. 5. Performance in both Forge Reply and Data Selfish (UDP)

(B) Misbehavior: Data Selfish

As shown in Figure 4, path-based reputation has improved throughput by 30% or more than the original AODV when the percentage of misbehaving nodes is under 40%, but the message overhead is only increased by less than 3%. (Delay results are omitted from now on since they are similar to those in Figure 3.) Thus, it is also quite effective in handling data selfish attacks. When compared with node-based reputation systems, its advantage is not as evident. We believe that is because this misbehavior is a "local behavior" that can be successfully detected by a node-based reputation system, whereas the forge reply attack affects the entire route, in which case the proposed path-based system is exceedingly useful.

(C) Misbehavior: Forge Reply plus Data Selfish

As shown in Figure 5, throughput is significantly improved when path reputation is used (more than double as compared to node reputation), yet the message overhead increases by less than 2% (delay is omitted since it is similar to the above). This again proves that the path-based reputation scheme is exceptionally effective at dealing with routing path disruption misbehaviors.

(D) Misbehavior: Data Modification

In this experiment, we did not see great improvement in throughput – about 20-30% when the rate of misbehavior is 20-40%; message overhead remains 4%. Results are not shown due to page limit. Again the results help one to see that the proposed system is not effective when dealing with "local" misbehaviors.

4.4   Simulation Results: FTP over TCP

In this set of experiments TCP traffic was applied with a message size of 512 bytes continuously sent from the ten randomly chosen sources to their corresponding receivers. The results are similar to those in UDP. Due to page limit, only the results of forge reply are shown.

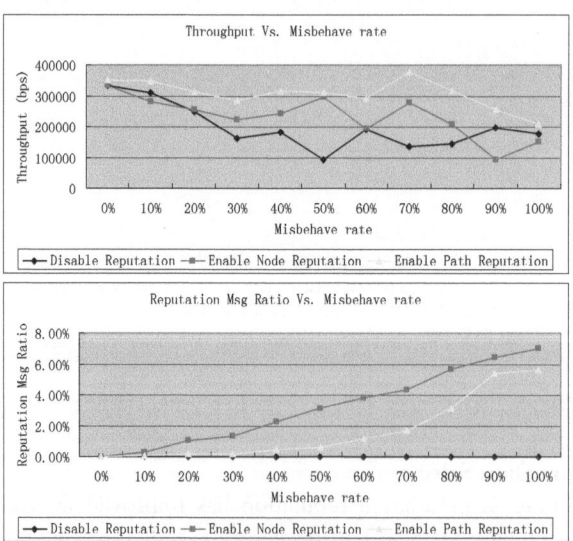

Fig. 6. Performance in Forge Reply (TCP)

Misbehavior: Forge reply
Yet, the results also show that the proposed reputation system was less efficient for TCP traffic than for UDP. We believe that this is mainly because the congestion control in TCP effectively limits its sending rate (and thus throughput) when misbehaviors take place. As the result, reputation systems are not as effective.

5 Conclusion

A path-based reputation system has been proposed that considers path reputation as a function of the reputation and trust of every node in the path. The system uses a category (or range) approach when evaluating the trust of second-hand information. It has been illustrated on top of AODV. Simulation results have shown that the proposed path-based system is most effective when handling routing misbehaviors such as forge reply (or worm-hole) attacks. In most cases, it doubles or even triples throughput. The extra message overhead is limited to 5%, which indicates that the reputation system is energy efficient. Future work may include modeling advanced adversary behaviors [1] and on/off misbehaviors in simulation, plus analyzing and fine tuning system parameters for optimal performance. In addition, it is possible to explore the use of fuzzy logic in the trust category approach we proposed, as it may be practical to fuzz the boarders between categories.

Acknowledgments

The authors would like to thank the anonymous reviewers for their comments and Ms. Diane Moh for her editorial help.

References

1. Anker, T., Dolev, D., Hod, B.: Cooperative and Reliable Packet Forwarding on top of AODV. In: Proceedings of the 4th International Symposium on Modeling and Optimization in Mobile, Ad-hoc, and Wireless Networks, Boston, MA, April 2006, pp. 1–10 (2006)
2. Bansal, S., Baker, M.: Observation-based cooperation enforcement in ad hoc networks. Technical Report CS/0307012, Stanford University (2003)
3. Buchegger, S., Le Boudec, J.Y.: Performance analysis of the CONFIDANT protocol. In: Proc. 3rd ACM Int. Symp. on Mobile Ad Hoc Network & comput., pp. 226–236 (2002)
4. Buchegger, S., Le Boudec, J.Y.: A Robust Reputation System for Mobile ad hoc Networks. EPFL IC_Tech_Report_200350 (2003)
5. He, Q., Wu, D., Khosla, P.: SORI: A secure and objective reputation-based incentive scheme for ad hoc networks. In: IEEE Wireless Communications and Networking Conference (WCNC 2004), Atlanta, GA, USA (March 2004)
6. Hu, J., Burmester, M.: LARS: a locally aware reputation system for mobile ad hoc networks. In: Proc. of the 44th Annual Southeast Regional Conference, pp. 119–123 (2006)
7. Li, J.: Design, Analysis, and Evaluation of Path-based Reputation System for MANET Routing, Master Thesis, Dept. of Computer Science, San Jose State Univ. (2008)

8. Li, X., Lyu, M.R., Liu, J.: A Trust model based routing protocol for secure ad hoc networks. In: Proc. of the IEEE Aerospace Conference, March 2004, pp. 1286–1295 (2004)
9. Lai, K., Baker, M., Marti, S., Giuli, T.: Mitigating routing Misbehavior in mobile Ad hoc networks. In: Proceedings of MOBICOM 2000, pp. 255–265 (2000)
10. Michiardi, P., Molva, R.: Core: a collaborative reputation mechanism to enforce node co-operation in mobile ad hoc networks. In: IFIP TC6/TC11 6th Joint Working Conf. on Comm. and Multimedia Security: Adv. Comm. and Multimedia Security, pp. 107–121 (2002)
11. Moh, M., Li, J.: A survey of reputation and trust systems for mobile ad-hoc network routing. Accepted to appear in Handbook of Communication and Information Security, edited by Stamp, M. Springer, Heidelberg (2009) (to be published in Fall 2009)
12. Ning, P., Sun, K.: How to misuse AODV: a case study of insider attacks against mobile ad-hoc routing protocols. In: Ad Hoc Networks, vol. 3(6), pp. 795–819 (November 2005)
13. Perkins, C.E., Royer, E.M.: Ad-hoc on-demand distance vector routing. In: Proc. 2nd IEEE Workshop on Mobile Computing Systems and Applications, pp. 90–100 (1999)
14. Raza, I., Hussain, S.A.: A Trust based Security Framework for Pure AODV Network. In: Proc. of the Int. Conf. on Information and Emerging Technologies (ICIET), Karachi, Pakistan, July 6-7, 2007, pp. 1–6 (2007)
15. Rebahi, Y., Mujica, V., Simons, C., Sisalem, D.: SAFE: Securing pAcket Forwarding in ad hoc nEtworks. In: Proc. 5th Workshop on Applications and Services in Wireless Networks (2005)
16. Zeng, X., Bagrodia, R., et al.: GloMoSim: a Library for Parallel Simulation of Large-scale Wireless Networks. In: 12th Int. Workshop on Parallel and Distributed Simulations (1998)

Hop-to-Hop Reliability in IP-Based Wireless Sensor Networks - A Cross-Layer Approach

Gerald Wagenknecht, Markus Anwander, and Torsten Braun

Institute of Computer Science and Applied Mathematics
University of Bern, Switzerland
{wagen,anwander,braun}@iam.unibe.ch

Abstract. To interconnect a wireless sensor network (WSN) to the
Internet, we propose to use TCP/IP as the standard protocol for all
network entities. We present a cross layer designed communication architecture, which contains a MAC protocol, IP, a new protocol called Hop-
to-Hop Reliability (H2HR) protocol, and the TCP Support for Sensor
Nodes (TSS) protocol. The MAC protocol implements the MAC layer of
beacon-less personal area networks (PANs) as defined in IEEE 802.15.4.
H2HR implements hop-to-hop reliability mechanisms. Two acknowledgment mechanisms, explicit and implicit ACK are supported. TSS optimizes using TCP in WSNs by implementing local retransmission of
TCP data packets, local TCP ACK regeneration, aggressive TCP ACK
recovery, congestion and flow control algorithms. We show that H2HR increases the performance of UDP, TCP, and RMST in WSNs significantly.
The throughput is increased and the packet loss ratio is decreased. As a
result, WSNs can be operated and managed using TCP/IP.

1 Introduction

Wireless sensor networks (WSN) consist of a large number of sensor nodes.
They are used for various applications, e.g., office buildings, environment control,
wild-life habitat monitoring, forest fire detection, industry automation, military,
security, and health-care. For such applications, a WSN cannot operate in complete isolation. It must be connected to an external network, e.g, the Internet.
Through such a network a WSN can be monitored and controlled. The operation
of a WSN needs a uniform communication protocol.

The TCP/IP protocol is the de facto standard protocol suite for wired communication. Using TCP/IP has a number of advantages and disadvantages. Thus, it
is possible to directly connect a WSN to a wired network infrastructure without
proxies or middle-boxes [1]. While in a TCP/IP supported WSN, UDP is used
to transmit sensor data to a sink, TCP would be used for administrative tasks
such as sensor configuration and code updates. Among the advantages, there
are a couple of difficulties running TCP/IP on sensor nodes. The resource constraints of sensor nodes and the high packet loss, which leads to a high number
of end-to-end retransmissions, result in a generally bad performance.

A couple of optimizations on different layers reduce the performance problems
when using TCP/IP. Avoiding the need of end-to-end retransmissions is the key

H. van den Berg et al. (Eds.): WWIC 2009, LNCS 5546, pp. 61–72, 2009.
© Springer-Verlag Berlin Heidelberg 2009

to increase the performance of TCP. This can be achieved by introducing hop-to-hop reliability mechanisms. UDP in WSNs benefits from hop-to-hop reliability as well. Furthermore, harmonizations between the protocols across different layers are an important target for optimizations. Thus, it is possible to achieve similar performance in terms of data throughput and packet loss rate with TCP/IP as if using common communication protocols for WSNs.

The following research questions arise. How can we design a protocol, which supports hop-to-hop reliability between two neighbor nodes? Can a cross-layer interface harmonize the different protocols on several network layers and increase the performance in terms of a better throughput (equivalent to lower transmission time) and lower error rate? Can TCP and UDP benefit from this architecture and run efficiently on sensor nodes?

The remainder of the paper is structured as follows. After the introduction in Section I, we present related work in the area of reliable transport protocols, TCP/IP adaptation, and cross-layer design for WSNs in Section II. In Section III we present the protocol stack and introduce our cross-layer interface. The Hop-to-Hop Reliability (H2HR) protocol, its cross-layer collaboration with the TCP Support for Sensor Nodes (TSS) [2] protocol and our beacon-less 802.15.4 MAC protocol [3] are described in Section IV. In the evaluation part in Section V, the implementation of the protocol using the OMNeT++ simulator is briefly described and the simulation results are presented. Section VI concludes the paper and gives an outlook.

2   Related Work

The use of TCP in wireless networks causes some serious performance problems [4], caused by end-to-end ACKs and retransmissions. A number of papers propose mechanisms to overcome these problems. In [5] the trade-off between TCP throughput and the amount of Forward Error Correction (FEC) is analyzed and simulated. In [6] the TCP performance is improved by establishing the optimal TCP window size. Caching and local retransmission are promising approaches to reduce the number of end-to-end retransmissions and make TCP feasible for WSNs. TCP Snoop [4] introduces first this approach. In [7] Distributed TCP Caching (DTC) is presented. TCP Support for Sensor Networks (TSS) [2] extends DTC by a novel congestion control mechanism that is very effective as well as easy to implement and deploy.

In the following, some common reliable transport protocols for WSNs are introduced. Directed Diffusion [8] is a popular data dissemination scheme. Reliable Multi-Segment Transport (RMST) [9] has been designed as a new transport layer for Directed Diffusion. It uses a NACK-based transport layer running over a selective-ARQ MAC layer to ensure reliability. It supports two modes: hop-to-hop and end-to-end mode. In the hop-to-hop mode it provides a caching mechanism on intermediate nodes. A lost packet is retransmitted from an intermediate node. In the end-to-end mode, lost packets are retransmitted from the source. Pump Slowly Fetch Frequently (PSFQ) [10] is also built on top of

Directed Diffusion. It runs over a non-ARQ MAC layer and ensures reliability by using sequence numbers and hop-to-hop recovery based on NACKs.

Besides ARQ, reliability ensuring link layer protocols have been developed for wired networks. Logical Link Control (LLC) is defined in IEEE 802.2. It provides a connection-oriented mode, which works with sequence numbers, and a connection-less mode with ACK frames. The LLC header includes a 16 bit control field and optionally the sequence number. The High-Level Data Link Control (HDLC) is bit-oriented and allows point-to-point and point-to-multipoint connections. Both protocols increase the complexity and overhead of the link layer significantly. The use of sequence numbers and additional header information waste space in the frames. The length of the frames defined in the 802.25.4 standard is limited to 128 bytes. Our H2HR protocol has no own header and does not use sequence numbers.

In [11], current activities in the area of cross-layer designs in WSNs are presented. Different cross-layer approaches are analyzed and a taxonomy to classify them is defined. Open challenges in the area of cross-layer designs are depicted.

3   Protocol Stack and Cross-Layer Interface

The protocol stack as shown in Fig. 1 includes the standard TCP/IP protocol suite with IP, TCP and UDP. The MAC protocol implements the beacon-less mode of the 802.15.4 MAC layer for peer-to-peer topologies. It supports two kinds of acknowledgment mechanisms, explicit ACK using ACK frames and implicit ACK using overhearing. The Internet Protocol remains unmodified. The Hop-to-Hop Reliability (H2HR) protocol is located between the Internet layer and the link layer. It increases the probability of successful delivery of a frame between two neighbor nodes, but does not guarantee reliability. It collaborates with the MAC protocol and the TSS protocol using the cross-layer interface. The UDP protocol is unmodified. Sensor data are transmitted using UDP from the sensor nodes to the base station. Although UDP is not a reliable transport protocol, it benefits from hop-to-hop reliability offered by the H2HR protocol. TCP is a protocol with end-to-end reliability, but it also benefits from H2HR in collaboration with TSS, which optimizes using TCP in WSNs. This is achieved by intermediate caching and local retransmission of TCP data packets, local TCP ACK regeneration, aggressive TCP ACK recovery, congestion and flow control algorithms.

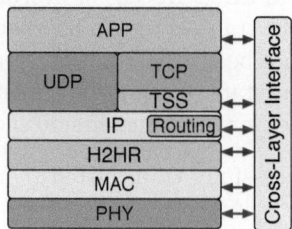

Fig. 1. Protocol Stack

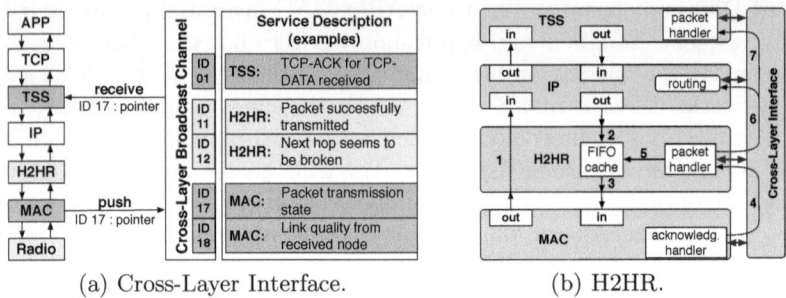

(a) Cross-Layer Interface.   (b) H2HR.

Fig. 2. Protocol Architecture

The cross-layer interface offers protocol interaction. Every protocol provides information that other protocols can use to optimize their operation. A cross-layer message consists of a unique ID and a pointer to the exchanged information. Fig. 2(a) shows the cross-layer interface and the interaction between the protocols.

There are two kinds of information exchange. First, a protocol subscribes to certain information. It uses the ID to identify the offered information uniquely. When an event has been registered, all subscribers are informed using the cross-layer broadcast channel. The second possibility for cross-layer information exchange is to request information directly from the protocols. The cross-layer interface gets the unique ID of the requested information and transmits the request to the according protocol using the cross-layer broadcast channel as well.

4 Hop-to-Hop Reliability Protocol and Cross-Layer Collaboration

The reliability between two neighbor nodes is ensured by the collaboration of the H2HR protocol with our beacon-less 802.15.4 MAC protocol and TSS across the layers. Fig. 2(b) shows the cross-layer collaboration between the protocols. The MAC protocol [3] supports two kinds of acknowledgment modes: explicit ACK using ACK frames and implicit ACK using overhearing. Both are 802.15.4 conform. In case of explicit ACK, the MAC protocol initiates the transmission of an ACK frame immediately after receiving the frame. The number of retransmission attempts is limited to three. In case of overhearing, no ACK frames are transmitted. Instead, the radio transceiver listens whether the next node forwards the frame. If a frame could not be overheard, it is retransmitted once. The upper layers are informed about the state of acknowledgments using the cross-layer interface. There are three states for the transmission success of a frame:

- The frame has been successfully transmitted (confirmed via ACK frame or overhearing).
- The frame has been transmitted, but there is no confirmation.
- Frame transmission failed.

The H2HR protocol is located between the IP layer and the link layer. Packets from layers below are delivered to the upper protocols according to the type of the packet without any processing (**1** in Fig. 2(b)). Packets from upper protocols are processed as follows. H2HR buffers the packets, which are delivered by IP (**2**) and delivers just one packet at a time to the underlying layers (MAC) (**3**). The other packets are buffered. The MAC layer initiates the transmission to the neighbor node. The H2HR protocol is informed about the transmission state by the MAC protocol using the cross-layer interface, either after the transmission has been successful or after three unsuccessful attempts (**4**). H2HR decides according to the state how the packet is handled (**5**). When the packet has successfully been transmitted to the next node, H2HR deletes this packet from the buffer and delivers the next packet to the underlying layer (**3**). Informed by the MAC protocol, H2HR reacts on two different kinds of problems. A packet can be lost due to interferences or due to congestion. If a packet has been transmitted, but there are no confirmations (neither ACK frame nor overhearing), the packet is lost due to interferences by a hidden node. In this case the transmission is retried immediately after 0.7-1.5 * frame_length (**3**). If the transmission has failed because the channel is busy, congestion is detected. H2HR initiates the retransmission after a random time between 1-2.5 * frame_length. The transmission of a 128 byte 802.15.4 MAC frame takes approximately 4ms with a data rate of 250kbps. After the 6th retransmission attempt initiated by H2HR, it can be expected that the neighbor node has serious problems or the channel is extremely busy. Thus, the packet is deleted and the routing protocol and TSS are informed using the cross-layer interface (**6**, **7**). Then, the routing protocol has to find an alternative route. The TSS protocol stores the packet and gets the control of the retransmission. After the route has been repaired or the channel is again free, retransmission is initiated by TSS.

UDP as an unreliable transport protocol can benefit from hop-to-hop reliability. The number of successfully transmitted packets from the sender to the receiver increases significantly. Overhearing with UDP is more challenging than using ACK frames, because UDP packets do not have any sequence numbers. The forwarded UDP packets can be identified by using a checksum over the UDP payload. In general, every transport protocol, which does not add sequence numbers to its packets, can use overhearing mechanisms at the link layer in this way.

TCP supports end-to-end reliability and lost packets are retransmitted by the sender. Because of the high packet loss ratio in WSNs, this happens quite often and pure TCP in WSNs is not feasible [4]. Hop-to-hop reliability supported by H2HR decreases the number of required end-to-end retransmissions, because it shifts the reliability assurance to intermediate nodes. Because H2HR can fail and does not guarantee the successful transmission of a frame to the next node, TCP end-to-end retransmissions can still occur. Thus, another protocol is required to support TCP in WSNs. TSS implements intermediate caching and local retransmission of TCP data packets, local TCP ACK regeneration, aggressive TCP ACK recovery, and congestion and flow control algorithms. If a packet is dropped by the H2HR protocol after 6 retransmission attempts due to busy channel, TSS still stores this packet in a buffer. If a TCP-ACK reaches the intermediate node and requests the presumably lost TCP data packet, then TSS transmits the cached packet to the receiver. The request for retransmission does not need to be transmitted to the sending side of the TCP connection. Finally, end-to-end retransmissions are only very rarely required and TCP can be used efficiently in WSNs.

As described above, ensuring reliability happens on different layers in different protocols. The MAC protocol reacts immediately on packet loss. If the frame transmission fails, H2HR intervenes and retransmits the packets depending on the detected problem (interferences or congestion). If H2HR collapses, the reliability mechanisms of the overlaying transport protocols handle the problem. This results in a hierarchy of reliability mechanisms.

H2HR does not require an own header and does not insert any information or sequence numbers into the frames. The length of the payload remains as large as possible. Furthermore, our protocols (802.15.4 MAC, H2HR, TSS) have low complexity and are easy to implement on sensor nodes, which have constraints in memory and processing power. Using the cross-layer interface, the protocols collaborate efficiently. A light-weight H2HR, a light-weight 802.15.4 MAC protocol and a light-weight TSS together have lower complexity as the combination of those mechanisms in one single protocol.

5   Evaluation

We implemented the MAC protocol, H2HR and TSS using the OMNeT++ simulator [12]. To compare a common transport protocol for WSNs with UDP and TCP, we implemented RMST in both modes. RMST is running over IP and 802.15.4, and not over Directed Diffusion [8] and 802.11 as proposed in [9]. Our implementation is based on the NS2 sources for RMST.

We analyze the influence of hop-to-hop reliability on the performance of UDP, TCP and RMST. We compare the packet loss ratio of UDP packets between the MAC protocol without reliability, implicit and explicit ACK, and the combination of H2HR with it. Afterwards, we compare the impact of the reliability mechanisms on the throughput using UDP, TCP with and without TSS, and RMST in the hop-to-hop and the end-to-end mode. Throughput is considered as the time required to transmit a certain number of bytes.

Four different scenarios are used to evaluate our cross-layer design communication architecture (*line scenario*, *cross scenario*, *parallel scenario*, and *grid scenario*, as shown in Fig. 3. To compare the transmission time, the paths in every scenario have 7 hops each. Data of 20 bytes or 1000 bytes are transmitted. In the line scenario, there is one connection $(0 \rightarrow 7)$. In the cross scenario, there are two connections $(0 \rightarrow 14, 1 \rightarrow 13)$. In the parallel scenario, there are two parallel connections $(0 \rightarrow 15, 1 \rightarrow 14)$. In the grid scenario, there are three connections, which end all on node 0. The connections are routed with the shortest path first algorithm. Every link is weighted equally. Thus, there are nodes $(9, 18)$, which hold two connections. We measure the transmission time of each connection separately.

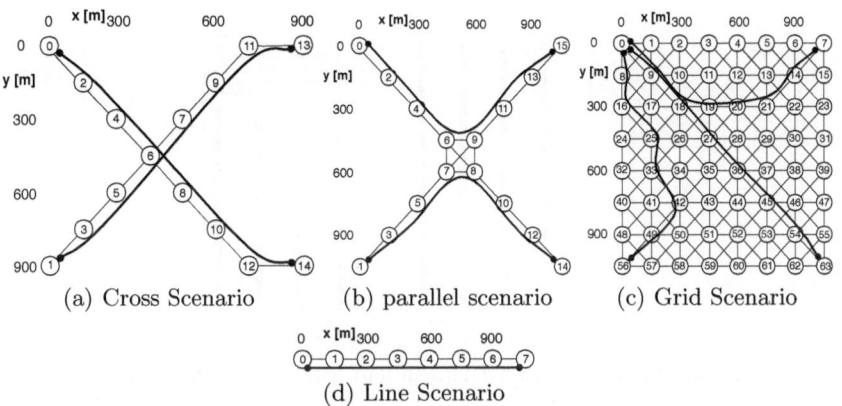

(a) Cross Scenario   (b) parallel scenario   (c) Grid Scenario

(d) Line Scenario

Fig. 3. Evaluation Scenarios

In all scenarios, no energy-saving functions such as duty cycles are implemented, because the focus is on the transmission performance.

For our simulation, we use an as realistic as possible radio model. According to the CC2420 manual [13] and the Castalia Simulator [14], we tuned the following values: carrierFrequency: 2.4E+9 Hz; bit-rate: 250 kbps; bandwidth: 10000 Hz; pathLossAlpha: 2; sensitivity: -95 dBm; thermalNoise: -110 dBm; dataLatency: 0.002 ms; CarrierSenseLevel: -77 dBm; transmission power: 1mW. The error rates between the nodes are very high, but reflect a real life scenario. In the line scenario, the average error rate between two neighbor nodes is around 20-25% according to previous measurements [3]. In the cross and square grid scenarios, the error rates are higher, because more nodes interfere with each other, especially in high traffic situations. The global buffer for the packets is limited to 5. In the explicit ACK mode, the MAC protocol has 3 retransmission attempts in the failure case. In the implicit ACK mode, a retransmission is retried twice. We use TCP Reno with a TCP window of 312 bytes. H2HR is configured as follows: the number of retransmission attempts is 6. Detecting a congestion, it waits chosen randomly between 4 ms and 11 ms. After detecting interferences H2HR, waits between 3 ms and 6 ms. Because the length of the MAC frame is limited to 128 bytes, the payload of a TCP packet is 78 bytes long and of a UDP packet is 90 bytes long. In the 20 bytes scenario, one packet is transmitted by UDP and RMST, and 5 packets by TCP. The 1000 bytes scenario requires 29 packets for TCP (including 13 data packets), 12 packets for UDP, and 11 packets for RMST. 50 different simulation runs were executed.

Fig. 4 shows the packet loss ratio of UDP packets transmitted in the four scenarios with a stream of 20 bytes and 1000 bytes. The percentage values are calculated over the absolute number of transmitted packets for all simulation runs. Without any reliability mechanism, the packet loss ratio is very high. When transmitting 1000 bytes, approximately 94% to 98% of packets are lost. When transmitting 20 bytes, the packet loss ratio is between 63% and 73%. One packet leads to less interferences. The packet loss ratio decreased when the simple MAC

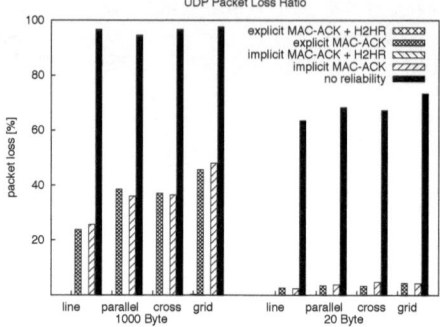

Fig. 4. Packet Loss Ratio for UDP with and without Reliability Mechanisms

reliability mechanisms without H2HR support is used. Using overhearing, the loss ratio is higher, because there are only two retransmission attempts. Due to less interferences between the nodes, the packet loss ratio in the line scenarios is lower than in the grid scenarios. Using H2HR in collaboration with the MAC protocol decreases the packet loss ratio dramatically. The probability of successful delivery of a packet is almost 100%, but there is no guaranteed reliability.

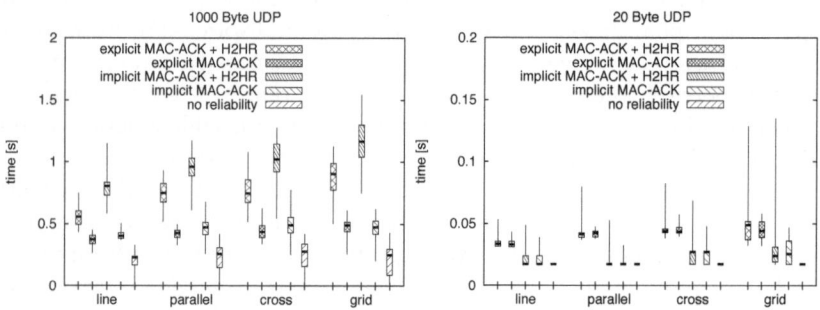

Fig. 5. Time Transmitting Data using UDP

Reliability mechanisms increase the probability of successfully delivered packets to the next hop, but it decreases the performance. Fig. 5 shows the influence of the reliability mechanism on the transmission time of 1000 bytes and 20 bytes respectively for the four different evaluation scenarios. Without reliability mechanisms, the required time to transmit the number of bytes is the shortest, but the packet loss ratio is very high. Just a few of the 12 packets of the 1000 bytes stream are successfully transmitted (usually just one or two). For example, the time required to reach the receiver is lower for packets number 3 and 8 than for packets number 2 and 12. In the 20 bytes scenarios, often no packet is successfully transmitted. In this case, the result is not used for the simulation results. The random backoff time of the MAC protocol has a small influence on the transmission time and adds some delay. Increasing the reliability with H2HR together with MAC retransmissions decreases the performance. Transmitting 1000 bytes takes between 550 ms (line scenario) and 1160 ms (grid scenario) compared to 370 ms up to 490 ms without any reliability mechanism. Using the MAC retransmissions without H2HR collaboration, the transmission time increases approximately 35% to 60%. The increased transmission time is caused by the retransmission attempts of H2HR, and the retransmission attempts of the MAC protocol. TCP tries to guarantee the delivery of packets. Without any additional reliability mechanisms, TCP does not work in typical WSNs. Even connection establishment does not work then. The TCP handshake needs two transmissions from sender to receiver and back (in our scenario this means 14 hops). The probability of a packet loss in the network is very high.

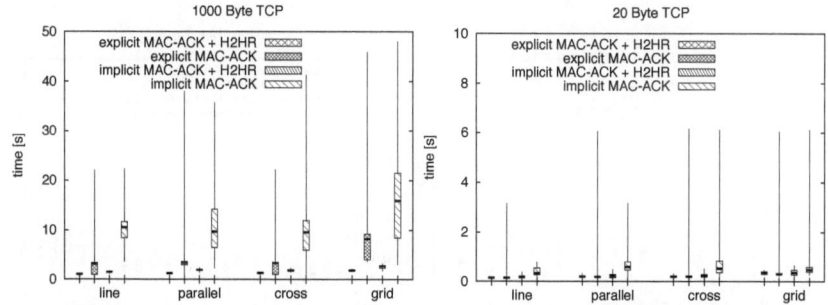

Fig. 6. Time Transmitting Data using Pure TCP without TSS

Fig. 6 shows the influence of H2HR on the transmission time for a TCP connection. We measured the time interval between the time a connection has been established and closed after the last packet has been transmitted successfully. We compared the influence of using H2HR with both ACK modes. In every scenario, transmitting 1000 bytes or 20 bytes using H2HR decreases the transmission time dramatically. In no case, a TCP end-to-end retransmission is necessary. The outliers in the boxplots without using H2HR occurred because of one or more TCP retransmissions. Each TCP retransmission costs approximately 3 seconds, which is the default value for a retransmission timer used by TCP Reno.

Fig. 7 shows the influence of having TSS as additional protocol. In general, using TCP with TSS is much faster than pure TCP. It optimizes the connection establishment. If H2HR is not used, it improves the performance by retransmitting lost packets within 1.5*RTT [2] by an intermediate node. Using H2HR with pure TCP and TCP with TSS has similar results. Using TSS smooths the outliers by preventing end-to-end retransmissions. Especially in scenarios with high traffic (1000 byte grid scenario for example), TSS has a positive influence on the transmission time. Because H2HR can react on interferences and congestion much faster than TSS or TCP, the influence of H2HR is much stronger. This can be seen in the outliers in Fig. 7(b). In these cases, the MAC protocol collapses and TSS intervenes (if H2HR is not active). If there are no problems, implicit ACK is faster then explicit ACK. No ACK frames have to be transmitted.

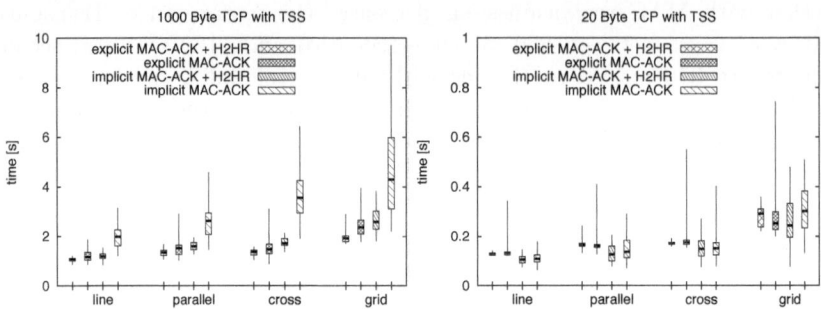

Fig. 7. Time Transmitting Data using TCP with TSS

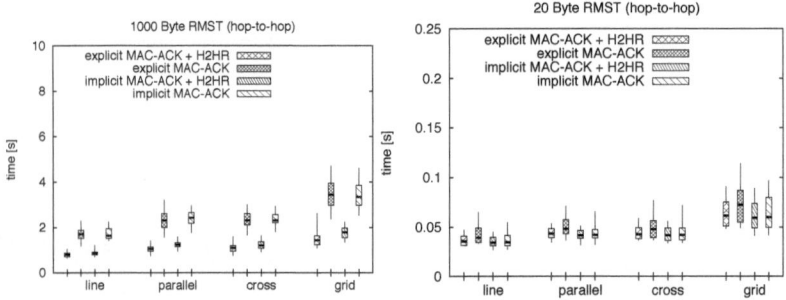

Fig. 8. Time Transmitting Data using RMST in the Hop-to-Hop Mode

Fig. 8 shows the measurements for the RMST protocol in the hop-to-hop mode. In the 20 bytes scenarios, there are almost no problems with interferences and congestion. H2HR has no influence, all problems are solved by the MAC protocol. The performance is almost as good as using UDP. It is much better than using TCP with TSS, because of additional packets for the connection establishment and the positive ACK frames. In the 1000 bytes scenarios, H2HR and the reliability mechanisms of RMST have to intervene. If H2HR is not active, RMST handles the retransmissions. This takes a long time, especially in the grid scenario with a lot of retransmissions caused by congestion. In the scenarios with less interferences and congestions, implicit ACK is faster than explicit ACK, because in the implicit ACK mode, no ACK frames have to be transmitted (the same has been observed for TCP with TSS, see Fig. 7). In the grid scenario, there are interferences and congestion. Node 9 and 18 hold 2 connections each. The third connection produces additional interferences (hidden node problem). Thus, packets have to be retransmitted (by the MAC protocol). The implicit ACK mode is slower than the explicit ACK mode, because the retransmission costs much more time (due to the random backoff time).

Fig. 9 shows the comparison of the several transport protocols (UDP, pure TCP, TCP with TSS, RMST end-to-end mode, and RMST hop-to-hop mode). H2HR is used in combination with explicit ACK mode. UDP has the best performance

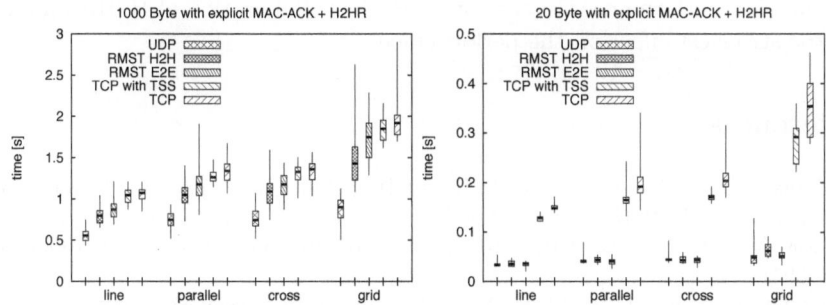

Fig. 9. Comparing the Transport Protocols using Explicit ACK and H2HR

in all scenarios. The reasons are clear: no connection establishment and no ACK frames are required. With H2HR, the successful delivery of all packets is very probable (see Fig. 4). But unlike to the reliable transport protocols, this is not guaranteed. Using UDP, there are outliers in the cross, parallel and grid scenarios. In these cases, the MAC protocol could not handle the situation (congestion, interferences) and H2HR retransmits the lost packets. RMST in both modes has a very good performance, because it uses negative ACKs and has no connection establishment. TCP with TSS performs better than pure TCP, because TSS prevents TCP end-to-end retransmissions (in case H2HR collapses) and optimizes the connection establishment. In the 20 bytes scenario, UDP and RMST have almost the same performance. The MAC protocol (or H2HR) can handle all situations and no retransmissions (end-to-end or hop-to-hop) are necessary in RMST. Because of the connection establishment and positive ACKs, the performance of pure TCP and TCP with TSS is significantly lower. In the 20 bytes scenario, there are less congestion situations and thus, the increased transmission time is caused by connection establishment. With TSS, the connection establishment and the acknowledgment handling is optimized. Generally we can say, with higher traffic (bigger packets to transmit), TCP with TSS has a similar performance as RMST. With lower traffic (just one packet to transmit), RMST is much better and performs similar to UDP. Bigger packets are typical for management tasks, e.g., code update.

6   Conclusion

In this paper we presented a cross-layer designed communication architecture containing a 802.15.4 conform beacon-less MAC protocol, the Hop-to-Hop Reliability (H2HR) protocol and TCP Support for Sensor Nodes (TSS). These protocols collaborate via a cross-layer interface. The H2HR protocol is harmonized with the beacon-less 802.15.4 MAC protocol and uses the acknowledgment mechanisms implemented by the MAC protocol. In general, hop-to-hop reliability mechanisms affect the performance of TCP, UDP, and RMST in WSNs. In case of UDP, H2HR increases the ratio of successfully delivered packets dramatically but at the expense of a higher transmission time. In case RMST, (in both modes), we showed that using H2HR produces the best results regarding the transmission performance. In case of TCP, the collaboration of H2HR and TSS has the strongest effect on the performance.

References

1. Dunkels, A., Voigt, T., Alonso, J., Ritter, H., Schiller, J.: Connecting Wireless Sensornets with TCP/IP Networks. In: Langendoerfer, P., Liu, M., Matta, I., Tsaoussidis, V. (eds.) WWIC 2004. LNCS, vol. 2957, pp. 143–152. Springer, Heidelberg (2004)
2. Braun, T., Voigt, T., Dunkels, A.: TCP Support for Sensor Networks. In: WONS 2007, Obergurgl, Austria, pp. 162–169 (January 2007)
3. Anwander, M., Wagenknecht, G., Braun, T.: Management of Wireless Sensor Networks using TCP/IP. In: IWSNE 2008, Santorini Island, Greece, pp. 1–8 (June 2008)
4. Balakrishnan, H., Seshan, S., Amir, E., Katz, R.H.: Improving TCP/IP Performance over Wireless Networks. In: Mobicom 1995, Berkeley, CA, USA, pp. 2–11 (November 1995)
5. Barakat, C., Altman, E.: Bandwidth Tradeoff between TCP and Link-level FEC. Computer Networks 39(2), 133–150 (2001)
6. Fu, Z., Zerfos, P., Luo, H., Lu, S., Zhang, L., Gerla, M.: The Impact of Multihop Wireless Channel on TCP Throughput and Loss. In: INFOCOM 2003, San Francisco, CA, USA, pp. 1744–1753 (April 2003)
7. Dunkels, A., Voigt, T., Alonso, J., Ritter, H.: Distributed TCP Caching for Wireless Sensor Networks. In: MedHocNet 2004, Bodrum, Turkey (June 2004)
8. Intanagonwiwat, C., Govindan, R., Estrin, D., Heidemann, J., Silva, F.: Directed Diffusion for Wireless Sensor Networking. IEEE/ACM Transaction on Networking 11(1), 2–16 (2002)
9. Stann, F., Heidemann, J.: RMST: Reliable Data Transport in Sensor Networks. In: SNPA 2003, Anchorage, AK, USA, May 2003, pp. 102–112 (2003)
10. Wan, C.Y., Campbell, A.T., Krishnamurthy, L.: PSFQ: A Reliable Transport Protocol for Wireless Sensor Networks. In: WSNA 2002, Atlanta, GA, USA, pp. 1–11 (September 2002)
11. Srivastava, V., Motani, M.: Cross-Layer Design: A Survey and the Road Ahead. IEEE Communications Magazine 43(12), 112–119 (2005)
12. OMNeT++: Discrete Event Simulation System, http://www.omnetpp.org
13. CC2420: Datasheet for the Chipcon CC2420 2.4 GHz IEEE 802.15.4 compliant RF Transceiver, Online (January 2009)
14. Pham, H.N., Pediaditakis, D., Boulis, A.: From Simulation to Real Deployments in WSN and Back. In: WoWMoM 2007, Helsinki, Finland, pp. 1–6 (June 2007)

A Scalable Security Framework for Reliable AmI Applications Based on Untrusted Sensors

José M. Moya, Juan Carlos Vallejo, Pedro Malagón,
Álvaro Araujo, Juan-Mariano de Goyeneche, and Octavio Nieto-Taladriz

Universidad Politécnica de Madrid, Dpto. Ingeniería Electrónica,
ETSI de Telecomunicación, Ciudad Universitaria s/n, 28040 Madrid, Spain

Abstract. Security in Ambient Intelligence (AmI) poses too many challenges due to the inherently insecure nature of wireless sensor nodes. However, there are two characteristics of these environments that can be used effectively to prevent, detect, and confine attacks: redundancy and continuous adaptation. In this article we propose a global strategy and a system architecture to cope with security issues in AmI applications at different levels. Unlike in previous approaches, we assume an individual wireless node is vulnerable.

We present an agent-based architecture with supporting services that is proven to be adequate to detect and confine common attacks. Decisions at different levels are supported by a trust-based framework with good and bad reputation feedback while maintaining resistance to bad-mouthing attacks. We also propose a set of services that can be used to handle identification, authentication, and authorization in intelligent ambients.

The resulting approach takes into account practical issues, such as resource limitation, bandwidth optimization, and scalability.

Keywords: Ambient intelligence, reputation system, security framework for wireless sensor networks.

1 Introduction

Security concerns have been identified as key issues in ambient intelligence (AmI) since its earliest inception (Weiser, 1993). Many in the research community clearly recognize the inherent challenge that an invisible, intuitive and pervasive system of networked computers holds for current social norms and values concerning privacy and surveillance. In fact, the increasing attack rate is becoming the bottleneck for adopting next-generation services and applications.

Three factors contribute to make security in intelligent environments even a much harder problem: 1) many nodes in the network have very limited resources; 2) pervasiveness implies that some nodes will be in non-controlled areas, and therefore potential intruders will have physical access to them; 3) all these computers are globally interconnected, allowing attacks to be propagated step by step from the more resource-constrained devices to the more secure servers with lots of private data.

H. van den Berg et al. (Eds.): WWIC 2009, LNCS 5546, pp. 73–84, 2009.

Usually, security issues are addressed, in a similar way to services in a network of general-purpose computers, by adding an authentication system and encrypted communications. But the resource limitations make the embedded computers especially vulnerable to common attacks.

In previous work [1], we demostrated that current ciphers and countermeasures often imply more resources (more computation requirements, more power consumption, specific integrated circuits with careful physical design, etc.), but usually this is not affordable for this kind of applications. But even if we impose strong requirements for any individual node to be connected to our network, it is virtually impossible to update hardware and software whenever a security flaw is found. It has already been stressed the need to consider security as a new dimension during the whole design process of embedded systems [2,3], and there are some initial efforts towards design methodologies to support security [4,5], but to the best of our knowledge no attempt has been made to exploit the special characteristics of AmI environments.

AmI applications have to live with the fact that privacy and integrity can not be preserved in every node of the network. This poses restrictions on the information a single node can manage, and also in the way the applications are designed and distributed in the network.

Of course, the inherent insecurity of embedded systems should not lead us to not try hard to avoid compromises. We should guarantee that a massive attack can not be fast enough to avoid the detection and recovery measures to be effective. Therefore we should design the nodes as secure as the available resources allow.

In spite of the disadvantages of AmI environments from the security point of view, there are two huge advantages of these kind of environments that we can and should use to fight against attacks:

- Redundancy. AmI environments usually have a high degree of spatial redundancy (many sensors that should provide coherent data), and temporal redundancy (habits, periodic behaviors, causal dependencies), and both can be used to detect and isolate faulty or compromised nodes in a very effective manner.
- Continuous adaptation. AmI environments are evolving continuously, there are continuous changes of functional requirements (data requests, service requests, user commands...), nodes appear and disappear continuously and therefore routing schemes change, low batteries force some functionality to be migrated to other nodes, etc.

In this article we propose a more secure approach to the design of AmI applications, by exploiting these two properties as much as possible. Section 2 describes our approach in detail. In section 3 we review some of the most relevant attacks and how this approach allows to detect and confine them. Finally, we draw some conclusions in section 4.

2   AMISEC Architecture

We focus on the development of secure applications in future wireless sensor networks, where many sensors provide data about observable magnitudes from the environment, and many actuators let the system act on the state of the environment.

Following the Ackoff taxonomy for the content of the human mind, we classify the content of the "ambient mind" into four categories:

1. Data: Symbols. It simply exists and has no significance beyond its existence (in and of itself).
2. Information: Data that is processed to be useful; provides answers to "who", "what", "where", and "when" questions.
3. Knowledge: Application of data and information; answers "how" questions.
4. Intelligence[1]: Appreciation of "why". It is the process by which new knowledge is synthesized from the previously held knowledge.

The main characteristic of an intelligent ambient is the semantic enrichment of environment based on the processing of data obtained from the environment using a sensor network. This "ambient mind" enhances the semantics of the environment by adding meaning to the objects. The objects are conscious of the "who", "what", "where", "when", "how", and "why".

Data is obtained by sensor nodes, but as they are not trusted, most of the remaining processing should be done in secure servers so that confidentiality attacks do not succeed (note that data has no meaning by itself). Data is sent to servers where it is processed to generate information, and then knowledge, and then understanding, and then new meaning, which is returned back to the environment. Individual nodes may be insecure, but the system should always continue its function of semantic enhancement. Moreover, attacks of individual nodes should not affect the decisions based on data from the environment. These requirements are achieved by perusing redundancy to discard data from the compromised nodes, and by changing the network structure and behavior at a speed that is fast enough to prevent a chained attack to spread.

As confidentiality attacks become more dangerous as data is further processed, there should be little or no processing at all in the sensor nodes, which are more vulnerable.

2.1 Network Model

We consider the network composed by two kinds of nodes: wireless nodes and servers. Wireless nodes provide data to the network to enable decisions to be made. They have to be inexpensive, what usually implies very limited resources, battery-powered, not maintained and hence insecure. Servers receive data from sensors and make decisions in order to reach the applications objectives. These decisions may imply to act in the environment and therefore they have to be secure. Servers are usually well maintained, wire-connected and their resources are not usually constrained.

---

[1] Actually, this category comprises two from the Ackoff taxonomy: understanding and wisdom.

In order to improve network scalability and throughput, we use a clustering technique based on Random Competition based Clustering (RCC) [6] to construct a multi-level network structures. Previous approaches [7,8,9] group nodes into clusters, and within each cluster a node is elected as a cluster head. Cluster heads together form a higher-level network, upon which clustering can again be applied. This structure simplifies communication and makes it possible to restrict bandwidth-consuming network attacks like flooding to a single cluster.

For a wireless network with n nodes capable of transmitting at $Wbits/s$, according to [10], the throughput, T, for each node under optimal conditions is

$$T = \Theta\left(\frac{W}{\sqrt{n}}\right)$$

Thanks to the clustering approach, in a two-level mobile backbone network where the number of nodes is n and the number of clusters is m, the throughput in the lower level becomes

$$T_1 = \Theta\left(\frac{W_1}{\sqrt{n/m}}\right)$$

and in the higher level

$$T_2 = \Theta\left(\frac{W_2}{\sqrt{m}}\right)$$

Node clustering, however, reduces redundancy and introduces single points of failure, as an intruder could control a whole zone by attacking its cluster head. The solution we propose is to introduce redundancy again. Every node in the network will have several cluster heads and will distribute messages randomly between them. This additional redundancy does not reduce the maximum throughput because at any given time the network structure is exactly the same as in the pure RCC scheme.

It may be argued that for every node to have two cluster heads, we need to double the backbone nodes so that there are twice as much backbone nodes in the coverage area. While it is true that more nodes have to belong to the backbone, this does not imply any reduction of the attainable throughput, as at any given time half the backbone nodes will not be used as such, and therefore the network structure remains exactly the same as in the pure RCC case. On the contrary, the burden of routing backbone messages is more distributed and therefore the penalty in energy consumption of being a cluster head is significantly reduced.

2.2   Assumptions

We assume that servers are secure and reliable.

The number of wireless nodes is assumed to be huge compared to the number of servers.

Due to being physically accessible and resource-constrained, wireless nodes are considered to be vulnerable. We assume an intruder can seize control of any wireless node in a minimum time ta.

As redundancy is good to detect and isolate attacks, any device providing useful information should be welcomed. Therefore, we assume that new wireless nodes can be added dynamically to our network without any restriction.

2.3   System Architecture

The AMISEC approach is based on leveraging the two weapons that we have to detect and resist to attacks and failures: redundancy (spatial and temporal), and continuous adaptation. Also, we know that individual wireless nodes are vulnerable to attacks, and therefore no important decision should be made by a single node and no significant information should be stored in a single node.

We propose a software architecture based on many independent agents with simple and clear responsibilities. An AMISEC agent is an independent piece of software that is able to act on your behalf while you are doing other things (they are proactive), and it does this based on its knowledge of your preferences and the context. This knowledge is stored in servers and it is available to the network nodes through the use of passive services.

Individual sensor nodes are not trusted by default, and therefore the notion of trust is built dynamically by comparing a sensor with its neighbourhood. For this reason, every agent that needs to take into account data coming from sensor nodes or any derived information uses a trust-based decision framework that is further described below.

Trust-based decision framework. We follow the definitions and beliefs of Boukerch et al. in [11] concerning the distinction between trust and reputation.

Trust is the degree of belief about the future behavior of other entities. Trust is subjective and it is based on past experiences.

Reputation, on the other hand, is the global perception of an entity's behavior, and it is based on the trust that others hold on that entity. It is mostly objective and it has some influence in the evolution of trust in every node.

To consider a data item to be valid we use two consistency tests. The data item is said to be s-consistent or consistent with the spatial redundancy if it is consistent with the data provided by the majority of sensors that provide measurements of the same variable. For example, for a presence event from a PIR detector to be valid, the majority of nodes monitoring the same area should also detect presence. In this evaluation every sensor is weighted with the trust value the receiving node has about the source node.

A second way to discard bad data is to evaluate each data item against temporal data redundancy. Each routing element stores a limited set of previous values for each variable directly routed through itself. The data item is said to be t-consistent if the variation against previous data is normal for that variable. For example, if a temperature value changes drastically and it is not maintained during some time, maybe a routing element has been attacked.

Both properties, s-consistency and t-consistency, are dependent on the variable being measured.

To model trust and reputation in our agent system, every node in the network maintains a trust table with entries for every relevant neighbor node.

When a new node is discovered, the initial trust value is 0.

Whenever a new message containing a new measurement of the external variable v arrives, trust on node i is recalculated as follows.

$$d_v(t) = A_v(\{\tau_i(t-1), d_{vi}(t)\})$$
$$\tau_i(t) = T(\tau_i(t-1), d_v(t), \overline{H_{vi}})$$

$\overline{H_{vi}}$ represents all the data values of the variable v provided by node i that are stored in this node (history is usually truncated to reduce memory requirements). A_v is an aggregation function that depends on the variable being measured, and it does not take into account data coming from a node with negative or zero trust value. T is also an aggregation function with these properties:

- If $\tau_i(t-1)$ is negative, the data item is discarded and no further processing is done for this message (repeated inconsistencies may lead to negative values of trust).
- If the new data element $d_{vi}(t)$ is s-inconsistent and t-inconsistent, it is stored in the local history (discarding the oldest value), but it is not taken into account for trust recalculation.
- If it is s-inconsistent with other sensors' data but t-consistent with previous values of the same sensor, trust on sensor i decreases.
- If it is s-consistent and t-consistent and current trust is positive, trust increases.

As can be seen, trust computation condenses historical information, and therefore it is bad, as we lose redundancy. On the other hand, resources are tightly constrained and we have to reduce storage requirements to a minimum.

To avoid some attacks, temporal disappearance means loss of positive trust (not negative). Whenever it appears again, it will get a 0 trust value.

There is a second method to feed trust values back from redundancy analysis: reputation messages from the servers.

From time to time, nodes communicate their trust tables to the servers. This is done at the routing level by adding this trust information to messages that are being sent to the same destination. Servers are not resource constrained by assumption, and therefore they can store all the historical information for future analysis. The adequate combination of all the trust data of a zone generates the global reputation data:

$$\rho_{vi}(t) = R(\rho_{vi}(t-1), H_{vi})$$

Where H_{vi} represents all the history of data values of the variable v provided by sensor i, and R is another aggregation function. Well-behaved nodes increase their reputation; bad-behaved ones decrease their reputation. Multiple agents can be running on the trust servers to look for attack evidences in the message history, and proactively reduce reputation values of suspect nodes.

Table 1. Parameters that can be adjusted dynamically to adapt the environment to possible attacks

Parameter Description	
Redundancy-related	
N_p	Number of reputation tables stored in a node.
N_d	Number of values stored for each sensor/value pair.
N_r	Number of routers per node.
Adaptation-related	
t_τ	Time between trust data messages sent to the reputation servers.
t_ρ	Time between reputation data messages from the servers to the nodes.
t_v	Time between sensor data messages from the sensor nodes to the network.
t_r	Minimum time between messages containing route information.

Whenever a server decides that it has to act in the environment by modifying trust values for ill-behaved nodes, it broadcasts the reputation information of all the nodes in that zone. This message is repeated from time to time until the data the server receives from that zone is consistent with the global reputation information.

A wireless node will never take into account this reputation information unless it has been received from different routers (cluster heads). Thus, redundancy in routing paths and trust merging in secure servers allows us to feed good and bad reputation back to the network without being vulnerable to bad mouthing attacks.

The trust data sent to the servers is enough to detect most, if not all, common attacks. However, it is not enough to find the concrete faulty or compromised node, and therefore the servers would not be able to confine the attack. The solution we propose is to include the routing path in some of the messages. This way, by analyzing the paths of messages with t-consistent and s-consistent data it is easy to discard well-behaved nodes. Note that routing paths coming from a compromised node could have been faked.

The confinement agents act directly by decreasing the reputation values of the suspect nodes.

A number of parameters (see table 1) can be dynamically adjusted in order to adapt the environment to possible attacks. If the risk increases, we increase the local amount of redundancy around the affected area.

Sensor agents are the simplest ones. They usually run on wireless nodes and provide measured data of external variables to the network, by sending messages to their routing agents. The message rate depends on the variation rate of the variable being monitored. This message rate should be enough to ensure that data items do not change too fast and therefore temporal redundancy can be used to detect failures or attacks.

Each sensor agent is associated to a sensor device and generates a sequence of measurements $d_{vi}(t), d_{vi}(t + 1), ...$ where v is the variable being measured and i is the sensor agent id. Each data item is annotated with a time stamp, to detect temporal anomalies.

Although they do not consume data from other sensors, they need to maintain a trust table for their routing elements, that will only evolve with reputation information coming from the servers. Unlike in routing elements, the initial trust value for a routing element is positive, and the distribution of messages is uniform between all the routing nodes with positive trust.

Actuator agents operate physically on the environment (light switches, electronic equipment controls, alarms, etc.). They are especially critical because 1) they are usually not redundant, and 2) any operation on them causes a physical effect on the environment. Therefore the nodes running actuator agents should be at least as tamper-resistant as the physical element they control. To ensure that an intruder can not operate remotely on an actuator, only servers can send operation requests to these agents and they should use robust asymmetric encryption algorithms. As security and processing requirements are higher, these nodes are usually main powered.

Data flows from sensors to servers and from servers to actuators. There is no feedback from actuators to servers. So if an actuator is attacked, the assailant will not be able to access to the others entities in the network.

Logically, an actuator works as a passive service, but it also develops a trust model of its environment, which is fed to the servers.

Aggregation agents reduce the redundancy by combining several data items using a known aggregation function. The only reason to apply these aggregations is to reduce the amount of data sent to the servers, allowing the system to scale.

Trust computation implies also an aggregation of spatial and temporal redundant data that is held in a node.

Services are passive elements that can be used by other nodes in the network. They usually run in servers.

Some of the services that have important roles for security reasons are: object tracking system, user tracking system, user modeling system, and common sense database.

2.4 Identity, Authentication, and Authorization

In this kind of environments there are two types of identity: object identity and user identity.

Objects are every traceable element in the network (a wireless sensor node, a camera, a remote control device, etc.). They are freely added to the network and they will only be isolated by the system in case of bad behavior.

Object identity is handled by the object tracking system, a server that stores and processes all the location information of network objects.

Different agents provide location information about the objects in the network.

From the security point of view, the main purpose of the object tracking system is to be able to detect and confine sybil attacks [12].

Authentication is implicit and linked to the concept of reputation. When the system has enough consistent data from an object identity, its reputation will grow and it is considered to be authenticated.

User identity is handled by the user tracking system. User identities are logical identifiers that are used to handle permissions in the environment. They are linked to objects automatically, based on the analysis of the data coming from the environment, the user model (preferences, habits, etc), and the common sense (we use a common sense database based on OpenMind's).

As previously seen, actuators have to be more secure because they can operate on the environment. No agent is allowed to use directly an actuator. They send an actuation request to the authorization service, and this service, if the object is linked to a user identity with permissions and the action is considered to be secure, will use the actuator. In our current implementation the authorization service holds the public keys for every actuator in the system and every operation message is encrypted with the public key of the actuator.

3   Evaluation

Nodes of a sensor network need to access, store, manipulate and communicate information. In AmI, nodes make decisions based on received data. Therefore, the system must guarantee data reliability. Some applications will require the use of sensitive information. In that case, measures to ensure data confidentiality should be taken into account. In this section, we will analyze the different kinds of attack that a sensor network is exposed to.

Confidentiality attacks attempt to access to the information stored in the sensor network. The network can use well-suited cipher algorithms [13] to provide security against attacks to communications. Due to resource constraints, nodes are more vulnerable to the attacks than communications. Some approaches suggest ciphering stored data [14]. Nevertheless, a combination of logical (cryptography weakness and Trojan horses), and physical (DPA, SPA, micro-probing, reverse engineering) attacks could break the ciphering and access the information.

Due to the characteristics of the sensor nodes, it is not possible to secure its data against attacks. Even if we cipher the information in the devices, an attacker could use an approach based on logical and physical attacks that could break the ciphering. Since attackers have physical access to the nodes and nodes have limited resources, vconfidentiality should be based in the main characteristics of sensor networks: distribution and redundancy.

Attacks against the confidentiality of node information attempt to access to the information stored in a sensor. If the attack successes, the attacker will obtain the information stored in it, but it is only raw data, not significant by itself. In addition, mapping that information with a concrete user is impossible because mapping information is stored in servers or distributed among a very large number of nodes. While the number of nodes holding some particular information remains much higher than the number of attacked nodes, attackers will not be able to obtain meaningful information.

When attacking to the confidentiality of communications, an intruder listens to the channel trying to obtain some information. Due to sensor redundancy and information distribution, the attacker should break all communications between sensors and routers to obtain some significant information. The use of some ciphering algorithms will help protecting the system. Since the network is big enough, an attacker that listens to the channel will obtain only a set of $d_{vi}(t)$. By definition, that set will not represent any meaningful information, so the attack will fail.

Whether it is jamming, collision or flooding, the effects of a denial-of-service attack in the network are similar: loss of messages and node disappearance. The seriousness and extension of the attack depends on the number of nodes, the stack layer where it takes place and several other parameters. Nevertheless, it leads some nodes to disappear. As no new value from these nodes arrives to the routers, as trust tables are sent to the servers, the global trust service will soon discover that the latest values coming from these nodes are obsolete and it will mark them as lost.

Local attacks can get worse if the compromised node stops routing properly, changes the values notified by some sensors, or teleports messages to other area of the network. A combined use of localization information (object tracking system), and route analysis for messages coming from the same area (redundancy in routing elements will ensure that not every message will go through the wormhole), allows to discover easily the bad routers. There are some proposals similar to this one, like in [15] where authors propose a method based on location information of each node and identity information in messages, or like in [16] where a statistical process of network data is used to detect wormholes. AMISEC manages the required data so both are feasible solutions for our system.

Integrity attacks are very difficult to avoid due to the weakness of wireless nodes. But these are clear cases of local attacks. Local or node attacks are not relevant for the AMISEC model, since redundancy allows losing nodes without any impact in the behavior. Negative reputation can be used from the servers in order to confine these attacks. Even if integrity of individual nodes is difficult to achieve, the use of redundancy can reduce or eliminate the impact on the global system.

Most identity attacks can be considered special cases of the Sybil attack [12]. This attack can be dangerous because it reduces the effect of the system redundancy. Our architecture avoids the Sybil attack by reducing its attack rate. When an aggregation agent receives information from an unknown node, the trust level default value is zero. This is enough to send data from this node to the servers to collect behavior history, but not enough to be taken into account in any decision or aggregation. If the node behaves correctly, its reputation will grow eventually, but always at a controlled rate. If many sensors are appearing in a short time in the same area, the required time to have positive reputation will increase.

4   Conclusion

Wireless Personal Area Networks are based on many wireless, low cost, low power, and low resources nodes. These characteristics and the possibility to access physically to the node make the nodes highly vulnerable to attacks. Cryptography appears as clearly insufficient to maintain data confidentiality and integrity in the network.

We have proposed a holistic solution that assumes this node vulnerability to address security issues in an intelligent ambient based on massive wireless sensor networks.

Redundancy and fast continuous adaptation have been identified as the key weapons to defend the system against attacks, and they are used consistently to cope with security issues at different levels.

The AMISEC architecture is based on an agent system with supporting services. Data flows from the sensors to the servers, where it is processed returning relevant semantic enhancements back to the environment. Agents running in insecure wireless nodes never hold a significant information unit, what preserves global confidentiality, and decisions are made in servers, what preserves integrity if redundancy is used adequately.

Most attacks are detected by the analysis of the redundant data available in the network and collected in the servers.

Decisions at different levels are supported by a trust-based framework where trust data only flows from the sensors to the servers and reputation only from the servers to the sensors.

The resulting approach takes into account practical issues, such as resource limitation, bandwidth optimization, and scalability.

Based on these results we claim that our approach provides a practical solution for developing secure AmI applications.

Acknowledgments. This work was funded partly by the Spanish Ministry of Industry, Tourism and Trade, under the CENIT Project Segur@, and partly by DGUI de la Comunidad Autónoma de Madrid and Universidad Politécnica de Madrid under Grant CCG07-UPM/TIC-1742.

References

1. Malagon, P., Vallejo, J., Moya, J.: Dynamic environment AmI evaluation for reliable AmI applications based on untrusted sensor. In: The International Conference on Emerging Security Information, Systems, and Technologies, 2007. SECUREWARE 2007, pp. 128–131 (2007)
2. Ravi, S., Raghunathan, A., Kocher, P., Hattangady, S.: Security in embedded systems: Design challenges. Trans. on Embedded Computing Sys. 3(3), 461–491 (2004)
3. Kocher, P., Lee, R., Mcgraw, G., Ravi, S.: Security as a new dimension in embedded system design. In: Ravi, S. (ed.) Proceedings of the 41st Design Automation Conference (DAC 2004), pp. 753–760. ACM Press, New York (2004)
4. Ravi, S., Raghunathan, A., Potlapally, N., Sankaradass, M.: System design methodologies for a wireless security processing platform. In: Proceedings of the 39th Conference on Design Automation, New Orleans, Louisiana, USA, June 10 - 14, 2002, pp. 777–782 (2002)

5. Arora, D., Raghunathan, A., Sankaradass, S.R.M., Jha, N.K., Chakradhar, S.T.: Software architecture exploration for high-performance security processing on a multiprocessor mobile soc. In: Proceedings of the 43rd Annual Conference on Design Automation, San Francisco, CA, USA, July 24-28, 2006, pp. 496–501 (2006)
6. Xu, K., Hong, X., Gerla, M.: Landmark routing in ad hoc networks with mobile backbones. In: Parallel Distributed Computing, pp. 110–122 (February 2003)
7. Bannerjee, S., Khuller, S.: A clustering scheme for hierarchical control in wireless networks. In. Proceedings of IEEE INFOCOM (2001)
8. Basagni, S.: Distributed clustering for ad hoc networks. In: Proceedings of the IEEE International Symposium on Parallel Architectures, Algorithms, and Networks, pp. 310–315 (June 1999)
9. Lin, C.R., Gerla, M.: Adaptive clustering for mobile wireless networks. IEEE Journal Selected Areas in Communications, 1265–1275 (September 1997)
10. Gupta, P., Kumar, P.: Capacity of wireless networks. Technical report, University of Illinois, Urbana-Champaign (1999)
11. Boukerch, A., Xu, L., EL-Khatib, K.: Trust-based security for wireless ad hoc and sensor networks. Comput. Commun., 11–12 (September 2007)
12. Douceur, J.R.: The sybil attack. In: Druschel, P., Kaashoek, M.F., Rowstron, A. (eds.) IPTPS 2002. LNCS, vol. 2429, pp. 251–260. Springer, Heidelberg (2002)
13. Mauro Conti, R.D.P., Mancini, L.V.: Ecce: Enhanced cooperative channel establishment for secure pair-wise communication in wireless sensor networks. Ad Hoc Networks 5, 49–62 (2007)
14. Subramanian, N., Yang, C., Zhang, W.: Securing distributed data storage and retrieval in sensor networks. Pervasive and Mobile Computing 3, 659–676 (2007)
15. ho Lee, K., Jeon, H., Kim, D.: Wormhole Detection Method based on Location in Wireless Ad-hoc Networks. In: New Technologies, Mobility and Security, pp. 361–372. Springer, Netherlands (2007)
16. Vajda, I., Buttyán, L., Dóra, L.: Statistical wormhole detection in sensor networks. In: Molva, R., Tsudik, G., Westhoff, D. (eds.) ESAS 2005. LNCS, vol. 3813, pp. 128–141. Springer, Heidelberg (2005)

The Quest for Mobility Models to Analyse Security in Mobile Ad Hoc Networks*

Mauro Conti[1,**], Roberto Di Pietro[2,***], Andrea Gabrielli[1],
Luigi Vincenzo Mancini[1], and Alessandro Mei[1,†]

[1] Dipartimento di Informatica, Università di Roma "Sapienza"
Via Salaria 113, 00198 - Roma, Italy
{conti,a.gabrielli,mancini,mei}@di.uniroma1.it
[2] UNESCO Chair in Data Privacy, Universitat Rovira i Virgili
Av. Països Catalans 26, 43700- Tarragona, Spain
roberto.dipietro@urv.cat

Abstract. Mobile Ad Hoc networks are subject to some unique security issues that could delay their diffusion. Several solutions have already been proposed to enforce specific security properties. However, mobility pattern nodes obey to can, on one hand, severely affect the quality of the security solutions that have been tested over "synthesized" mobility pattern. On the other hand, specific mobility patterns could be leveraged to design specific protocols that could outperform existing solutions.

In this work, we investigate the influence of a realistic mobility scenario over a benchmark mobility model (Random Waypoint Mobility Model), using as underlying protocol a recent solution introduced for the detection of compromised nodes. Extensive simulations show the quality of the underlying protocol. However, the main contribution is to show the relevance of the mobility model over the achieved performances, stressing out that in mobile ad-hoc networks the quality of the solution provided is satisfactory only when it can be adapted to the nodes underlying mobility model.

Keywords: wireless Ad Hoc networks security, mobility models, node capture attack detection, distributed protocol, resilience.

1 Introduction

The capacity of Ad Hoc networks to operate without requiring an existing infrastructure, makes Ad Hoc networks an ideal candidate for the deployment in harsh environments to fulfill law enforcement, search-and-rescue, disaster recovery, and other civil

---

* This work was partly supported by: The Spanish Ministry of Science and Education through projects TSI2007-65406-C03-01 "E-AEGIS" and CONSOLIDER CSD2007- 00004 "ARES", and by the Government of Catalonia under grant 2005 SGR 00446.
** Corresponding author.
*** Also with Dipartimento di Matematica, Università di Roma Tre, L.go S. Leonardo Murialdo 1, 00146 - Roma, Italy. E-mail: dipietro@mat.uniroma3.it
† Alessandro Mei was partially funded by the FP7 EU project "SENSEI, Integrating the Physical with the Digital World of the Network of the Future", Grant Agreement Number 215923, www.ict-sensei.org

H. van den Berg et al. (Eds.): WWIC 2009, LNCS 5546, pp. 85–96, 2009.
© Springer-Verlag Berlin Heidelberg 2009

applications. In these cases, another appealing operating feature is for Ad Hoc networks to operate in unattended manner. However, this comes at a cost: Ad Hoc networks can be prone to different kinds of novel attacks. For instance, an adversary could eavesdrop all the network communications, or it might capture (i.e. remove) nodes from the network. These nodes can then be re-programmed and deployed within the network area, with the goal of subverting the data aggregation or the decision making process. Another range of possible attacks is known as *sybil attack* [17], where a single node illegitimately claims multiple identities —stolen from previously captured nodes. Finally, the *clone* attack occurs when a node is first captured, then tampered with, re-programmed, and finally replicated in the network. A few techniques exist to delve with the former attack: based on RSSI [7]; leveraging key-based authentication; and, probabilistic solutions based on node cooperation [4].

An application for node capture detection could be the LANdroids [13] research program by the U.S. Defense Advanced Research Projects Agency (DARPA). This research program has the aim to develop a so-called: Smart robotic radio relay nodes for battlefield deployment. LANdroid mobile nodes are supposed to be deployed in hostile environment, establish an ad-hoc network, and provide connectivity as well as valuable information for soldiers that would later approach the deployment area. An adversary might attempt to capture one of these nodes. We are not interested in the goals of the capture (that could be, for instance, to re-program the node to infiltrate the network, or simply extracting the information stored in it); but on the open problem of how to detect the node capture that represents, as shown by the above cited examples, a possible first step to jeopardize an Ad Hoc network. Indeed, an adversary has often to capture a node before being able to launch other more vicious, and may be still unknown, attacks. Node capture is one of the most vexing problems in Ad Hoc network security [18]. In fact, it is a very powerful and hard to detect attack. We believe that any solution to this problem has to meet the following requirements: (i) to detect the node capture as early as possible; (ii) to have a low rate of false positives—nodes that are believed to be captured and thus subject to a revocation process, but that were not actually taken by the adversary; (iii) to introduce a small overhead. The solutions proposed so far are not efficient [18]. Moreover, due to the distributed nature of Ad Hoc networks, naïve centralized solutions, although it can be in principle applied, present drawbacks like single point of failure and non uniform energy consumption. The unique requirements of the Ad Hoc network context call for efficient and distributed solutions to the node capture attack.

The contribution of this work is to investigate the influence of a realistic mobility scenario over our proposed framework. In particular, we consider the traces in [20]: The traces were collected distributing iMotes to students attending the 3-days INFOCOM 2005 student workshop. The number of devices is 41, and they were programmed to log contacts of the meeting with other devices. We study the characterization of these real traces against the Random Waypoint Mobility Model. Extensive simulations results show the quality of the underlying protocol for the node capture detection as well as the fact that mobility models have a relevant impact on the performance of the underlying algorithm. This provides an insight on the fact that protocols design cannot be separated from the underlying mobility models they are supposed to operate upon.

The paper is organized as follows. Section 2 presents the related work in the area. Section 3 describes our approach for the node capture detection in Mobile Ad Hoc networks. In Section 4 we report the results of an extensive set of simulations of our proposed approach considering both the RWMM and a real mobility trace. Finally, Section 5 introduces some concluding remarks.

2   Related Work and Background

Mobility as a means to enforce security in mobile networks has been considered in [3]. Further, mobility has been considered in the context of routing [10] and of network property optimization [15]. In particular, [10] leverages node mobility in order to disseminate information about destination location without incurring any communication overhead. In [15] the sink mobility is leveraged to optimize the energy consumption of the whole network. A mobility-based solution for detecting the sybil attack has been recently presented in [19]. Finally, note that a few solutions exist for node failure detection in Ad Hoc networks [11]. However, such solutions assume a static network, missing a fundamental component of our scenario, as shown in the following.

We recently proposed a solution that uses node mobility to cope with the node capture attack [5]. The solution relies on the meeting frequencies between honest nodes to gather information about the absence of captured nodes. A property similar to that of node "re-meeting" has been already considered in [6]. However, in [6], the authors investigate the time needed for a node to meet (for the first time) a fixed number of other nodes. While the results given in [5] are encouraging, the specific solution proposed requires a high overhead to bound the number of false positive (wrongly revoked nodes). Note that, without this bounding mechanism the number of false positive would be unacceptable. In this work we consider a slightly improved version than the one in [5]. In particular, the improvement is a a presence-proving mechanism used by allegedly captured nodes to show their actual presence in the network (that is, bounding the number of false positive).

Our previous solution [5] is based on the simple observation that if node a will not *re-meet* node b within a period λ, than it is possible that node b has been captured. The solution is build upon this intuition to provide a protocol that makes use of local cooperation and mobility to locally decide, with a certain probability, whether a node has been captured or not. The solution does not rely on any specific routing protocol: We resort to one-hop communications and to a sparing use of a message broadcasting primitive. This distinguished feature helps keep our protocol simple, efficient, and practically deployable, avoiding the use of sophisticated routing that can introduce complexity and overhead in the mobile setting. Furthermore, our experimental results demonstrated the effectiveness and the efficiency of our proposal. For instance, for a given energy budget, while the benchmark requires about 4,000 seconds to detect node capture, our proposal requires less than 2,000 seconds.

The previous work [5] studied the proposed capture detection solution considering the Random Waypoint Mobility Model (RWMM). While the RWMM has been widely used in literature [23, 2] different drawbacks of this model has been raised [12, 24], and it might not be the best model to capture a "realistic" mobility scenario, as highlighted in [21]. In the last two years, different interesting mobility model has been proposed, particularly for human mobility [16]. However, while a mobility model is useful to fast and easily analyse new protocol ideas, the best way to understand how a protocol performs on real scenario is to use real mobility traces.

3   Leveraging Node Mobility and Node Cooperation for the Capture Detection

The aim of a capture detection protocol is early detection of a node removal. In the following we also refer to this event as a node capture. The capture detection protocol should be able to identify which is the captured node, so that its ID can be revoked from the network. Revocation is a fundamental feature—if the adversary reintroduces the captured (and possibly reprogrammed) node in the network, the node should not be able to take part to the network operations.

Our node capture detection approach [5] is based on the intuition that leveraging node mobility and cooperation helps node capture detection. In this section we present an improved version of the protocol in [5]. In particular, Section 3.1 presents the assumptions and notations, Section 3.2 presents an overview the protocol while we use the improved version for studying the performances in real mobility traces in Section 4.

3.1   Assumptions and Notation

The assumption at the base of our protocol is that if a node has been absent from the network for a given interval time (that is no one can prove its presence in that interval) the node has been captured. Indeed, we could incur in wrong revocation if the node is actually not captured but, for example, only disconnected for that considered time interval.

In the following we define a *false positive alarm* as an alarm raised for a node that is actually present. Another issue is Denial of Service (DoS). Indeed, since alarms are flooded in the network (as it will be clear in the following), it could be possible for a corrupted node to trigger false alarms so as to generate a DoS. This issue is out of the scope of this paper, however, for the sake of completeness, we sketch in the following a possible solution. The impact of false positives can be mitigated noting that it could be possible, once the recovery mechanism detects a false alarm, to associate a failure tally to the node that raised the false alarm. If the tally exceeds a certain threshold, the appropriate action to isolate the misbehaving node could be take.

Further, we assume the existence of a failure-free node broadcasting mechanism [14]; and, finally, we point out that addressing node-to-node secure communications,

Table 1. Time-related notation

Symbol	Meaning
σ	Message propagation delay.
λ	Alarm time used in the cooperative protocol.
δ	Time available to the allegedly captured node to prove its presence.

addressing confidentiality, integrity, privacy, and authentication are out of the scope of this paper. However, note that a few solutions explicitly addressing these points can be found in literature [22, 8].

Table 1 resumes the intervals time notation used in this paper.

3.2 The Protocol

In this section we describe an improved version of our previously proposed detection protocol [5]: The node Capture detection protocol that leverages Mobility and Cooperation (in the following also called CMC Protocol). We start from the following observation: If node a has listened to a transmission originated by node b, at time t, we will say that a *meeting* occurred. Now, nodes a and b are mobile, so they will leave the communication range of each other after some time. However, we expect these two nodes to re-meet again within a certain interval of time, or at least within a certain time interval with a certain probability. Basically, each node a is given the task of witnessing for the presence of a specific set T_a of other nodes (we will say that a is *tracking* nodes in T_a). For each node $b \in T_a$ that a gets into the communication range of, a sets the corresponding meeting time to the value of its internal clock and starts the corresponding time-out, that would expire after λ seconds. The meeting nodes can also cooperate, exchanging information on the meeting time of nodes of interests —that is, nodes that are tracked by both a and b. Note that node cooperation is an option that can be enabled or disabled in our protocol. If the time-out expires (that is, a and b did not re-meet within λ seconds), the network is flooded with an alarm triggered by node a. If node b does not prove its presence within δ seconds after the broadcasted alarm is flooded, every node in the network will revoke node b.

The CMC protocol is event based. In particular, every node a has to manage the following events:

- Meeting with node b (Procedure *CMC_Meeting*);
- A alarm time-out (λ) expires (Procedure *CMC_TimeOut(ID$_x$)*);
- Receiving a message m; (Procedure *CMC_Receive(m)*)

In the following, we briefly describe the CMC protocol operations for each of the previous three steps. We refer to [5] for a deeper description of the protocol.

3.3 Meeting

When node a meets node b, node a checks if it is supposed to trace node b (that is if $b \in T_a$). Assume that $b \in T_a$, then a further check if b is already revoked: To do so, each node stores a Revocation Table (RT_a) that lists the revoked nodes. If both previous tests succeed, then a updates the information about the last meeting with node b. Node a stores in a Time-out Table TT_a the ALARM and REVOKE time-outs (the use is explained in Section 3.4).

In the cooperative version of the protocol the following scenario is also considered. Assume nodes a and b are both tracing a third node c. Assume the last meeting between a and c happened at time t_1 and the last meeting between b and c happened at time t_2. If $t_1 > t_2$ ($t_2 > t_1$), i.e. a met c more recently than b, the cooperative algorithm also executes the *CMC_Meeting* procedure on node a (b), considering a meeting with node c. We refer to this non-physical meeting as *virtual* meeting.

When two node meets they send each other a CLAIM message (something similar is simulated for *virtual* meetings). The use of ALARM and CLAIM messages is explaned in Section 3.5.

3.4 Time-Out Expires

If an ALARM time-out Expires—node a did not meet node ID_b for a time λ—a floods the network with an ALARM message and set a new REVOKE time-out for node b.

If a REVOKE alarm expires—δ seconds elapsed from the alarm triggering without suspected captured node presence evidence—a node revocation procedure for the suspected node is executed.

3.5 Receiving a Message

Assume the message is of type ALARM. The executing node checks that the alarm is not related to itself and the suspected node, ID_x, is not already revoked. In this case, the REVOKE time-out is set for node b. If the ALARM is related to the executing node itself, node a floods the network with a presence CLAIM message. This measure prevents *false positive detection*—that is, the revocation of nodes that are active in the network.

On the other hand, assume the received message is of type CLAIM (this means that a node that was the target of an ALARM message is proving its presence). A *virtual* meeting between a and the wrongly accused nodes is triggered. The overall result is that node a disables the REVOKE time-out for that node while restarting the ALARM time-out for the same node.

4 Simulations and Discussion

In this section, we present the results of an extensive set of simulations of our capture detection protocol. In particular, we show simulations results analyzing the characterization of real mobility and the main difference with the RWMM.

While mobility models based on randomly moving nodes may, for example, provide useful analytical approximations to the motion of vehicles that operate in dispatch mode or delivery mode [1], it is not directly applicable to others scenario-inspired mobility models [21]. In this section, we present the results of the simulations that we run in order to understand the characterization of a real mobility pattern. We run these simulations for both the INFOCOM mobility trace and the RWMM in order to also get the difference between a synthetic model compared to a realistic one.

We implemented a simulator of our protocol that takes as input a trace of nodes mobility—every nodes meeting is described by the couple of participating nodes id and the time of the meeting. We ran multiple simulations of the CMC protocol (with and without cooperation) on both the INFOCOM trace and the RWMM synthetic trace. We generate the synthetic trace for the RWMM using the previous implementation of the RWMM [5], considering 41 nodes deployed in a $1,000 \times 1,000 m^2$ area and a transmission

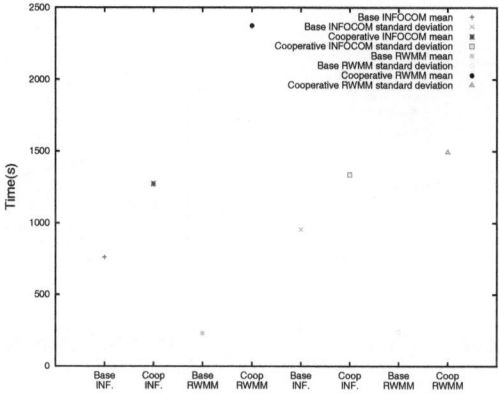

Fig. 1. Mean and Standard Deviation of the Detection Time: $n=41$, $|T|=41$, $\lambda=5,400s$, $\delta=30s$

range of 15 meters. Note that, for generating the RWMM synthetic trace, we set the parameters values similar to the ones used for the INFOCOM trace [20] ([20] considers 41 Bluetooth devices, where radius nodes range from 10 to 20 meters).

We considered the events of the first day split in 8 intervals of 3 hours each. For each of the 41 nodes, we simulated its capture at the end of every given intervals; these resulting in a total of 328 simulations for every combination of protocol version (base and cooperative) and trace (RWMM and INFOCOM). For every point we run the protocol and measured the detection time. For every combination setting (version and trace), Figures 1 reports the mean and the standard deviation of the detection time for all of the 41*8=328 points. We observe that for both the RWMM and the INFOCOM trace, the means in the base version of the protocols are lower than in the cooperative protocol. However, we cannot conclude that the base version is better than the cooperative because of energy consumption aspects later described. If we compare the means on the INFOCOM trace with the means on the RWMM trace, we observe that the means of the base and cooperative protocols on INFOCOM have a mean detection time lower than the cooperative in the RWMM, but higher than the base in the RWMM.

To better explain the increase of the detection times in the cooperative version, we plot the number of false alarms sent by the nodes when the simulations are run without capturing any node—this is a protocol overhead cost. In particular, Figure 2 shows for every node in the network (x-axis), the number of false alarms (z-axis) it sends claiming the capture of another node (y-axis).

From Figure 2, we observe that the base protocol in the RWMM (Figure 2(a)) has a high number of alarms, while the cooperative protocol on the same trace (Figure 2(b)) has a low number of alarms. The same pattern is present in the Figure 2(c), where the number of false alarms in the base is higher then in the cooperative (Figure 2(d)). Note that, the base protocol relies only on its own information over the monitored nodes. On the other hand, the cooperative version uses also evidences collected by other nodes. This difference makes the cooperative version more accurate than the base version on the raise of alarms. In fact, the number of false alarms sent by each node is higher in

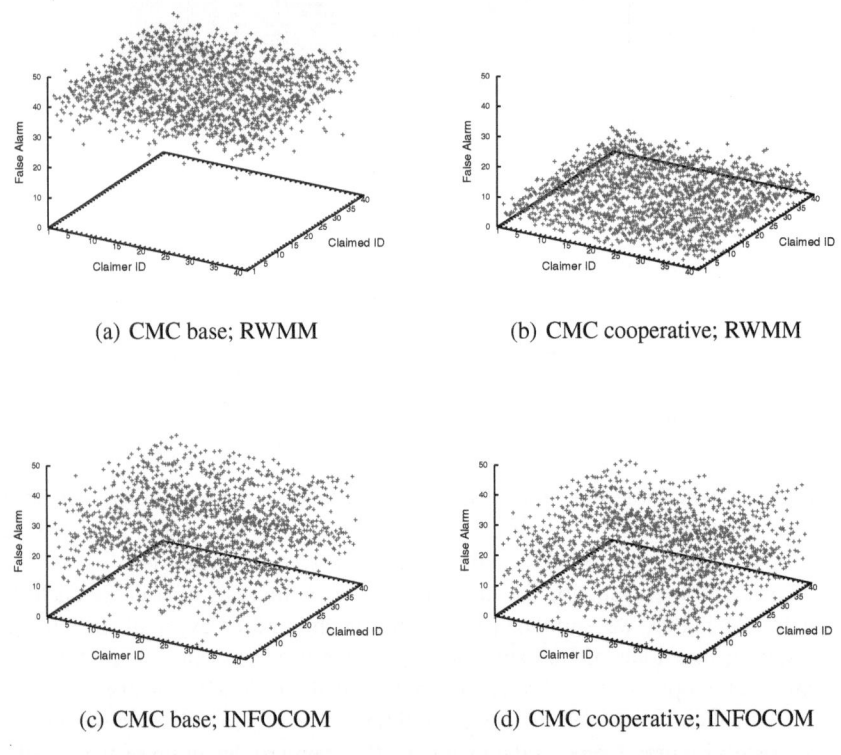

(a) CMC base; RWMM   (b) CMC cooperative; RWMM

(c) CMC base; INFOCOM   (d) CMC cooperative; INFOCOM

Fig. 2. Number of false alarms: $n=41$, $|T|=41$, $\lambda=5,400s$, $\delta=30s$

the base version rather that in the cooperative one. As a consequence, the detection time in the base protocol is lower than in the cooperative one because of the higher rate of alarm raised (this motivates the shape of Figures 1). Note that, the false alarms of the RWMM graphs are grouped together; on the top of the graph for the base protocol and on the bottom for the cooperative. Another observation that can be expressed, is that the false alarms for the INFOCOM trace are more distributed, and they present some isolated points, as node 31. This is explained because of the peculiarity of the real mobility patterns versus the synthesized one. In fact, the synthesized mobility patterns tends to uniform the nodes' behavior.

We plot Figure 3 to better compare the difference on the number of false alarms raised in the RWMM trace and in the INFOCOM trace. Figure 3 summarized the results of Figure 2. In particular, Figure 3 shows for each node in the network (x-axis) the number of alarm its sends (y-axis).

Figure 4 shows for every node in the network (x-axis), the number of meetings (y-axis). We observe that the synthesized trace (Figure 4(a)) makes the number of meetings uniform. In particular, all the nodes have a number of meeting close to 580. On the other hand, in the real mobility trace, we have both nodes with a high number of meetings, i.e. node 40 has about 1,400 meeting, and nodes with a low number of meeting, i.e. nodes 12 and 31 have less than 200 meetings.

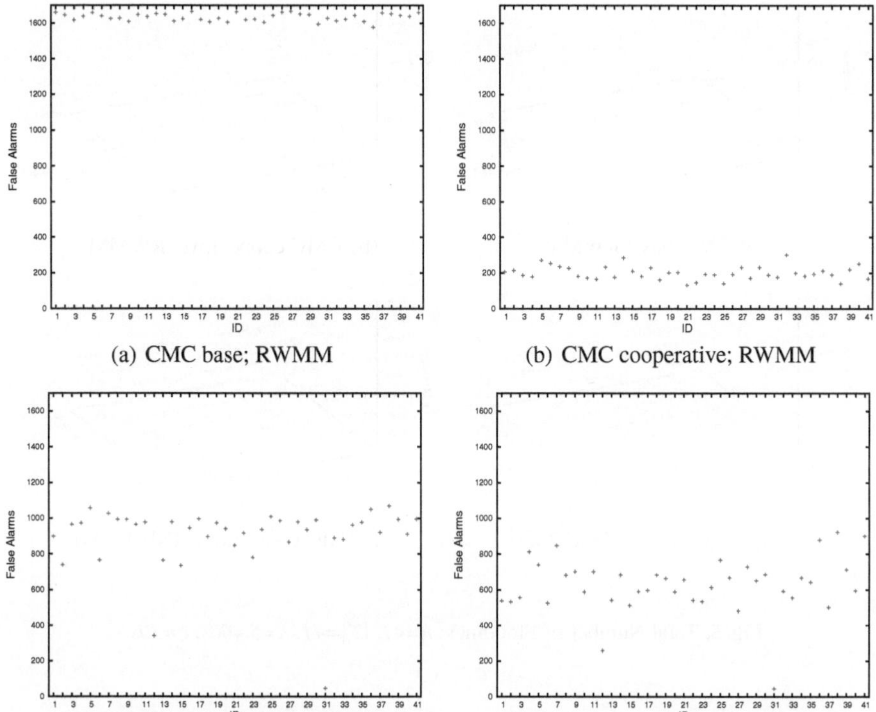

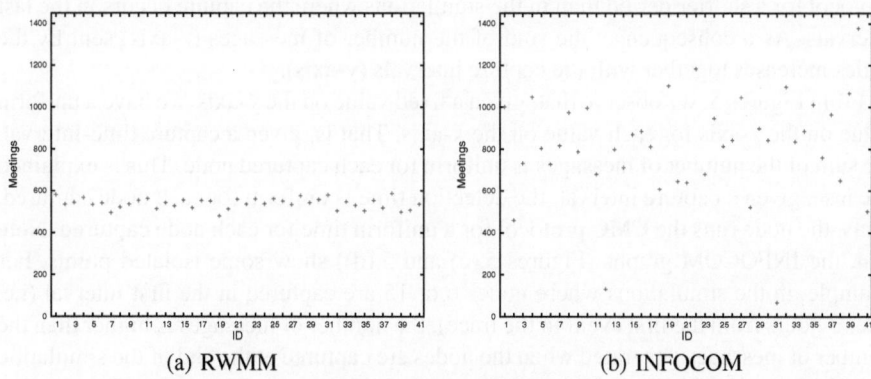

Fig. 3. Sums of false alarms: $n=41$, $|T|=41$, $\lambda=5,400s$, $\delta=30s$

(a) CMC base; RWMM

(b) CMC cooperative; RWMM

(c) CMC base; INFOCOM

(d) CMC cooperative; INFOCOM

(a) RWMM

(b) INFOCOM

Fig. 4. Total Number of Meetings

In Figures 5, we plot the sums of the number of messages sent by all the nodes for everyone of the 328 previous simulations. In particular, for each node (x-axis) and for each of the 8 capture intervals (y-axis) we plot the sums of the number of messages sent (z-axis) by all the nodes in the network. Each simulation ends when the capture of the

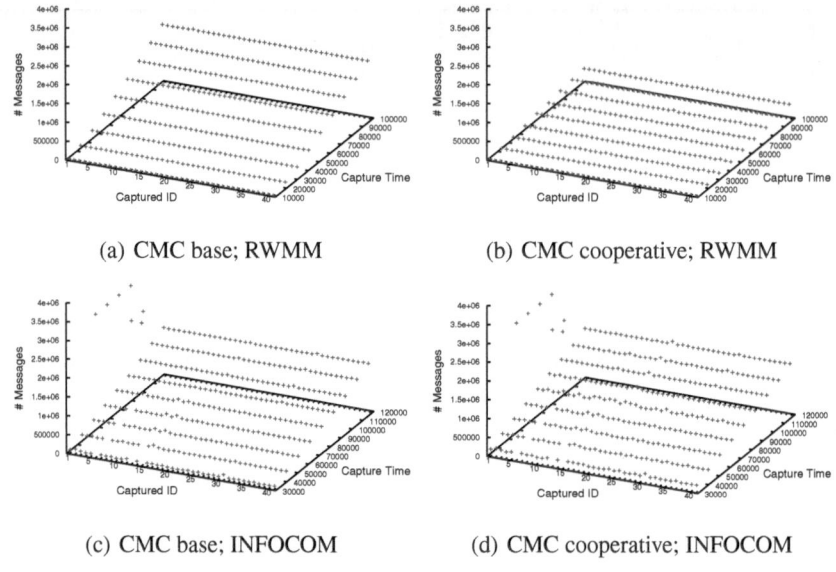

(a) CMC base; RWMM   (b) CMC cooperative; RWMM

(c) CMC base; INFOCOM   (d) CMC cooperative; INFOCOM

Fig. 5. Total Number of Floodings: $n=41$, $|T|=41$, $\lambda=5,400s$, $\delta=30s$

node is detected. If a node is captured in the first intervals (i.e. after 3 hours from the first event in the trace) the detection will occur in a shorter time than if it is captured towards the final intervals (i.e. after 24 hours from the first event in the trace). Thus, in the simulations where the capture occurs in the first intervals, the nodes run the CMC protocol for a shorter period than in the simulations where the capture occurs in the last intervals. As a consequence, the sum of the number of messages (z-axis) sent by the nodes increases together with the capture intervals (y-axis).

From Figures 5, we observe that, given a fixed value on the y-axis, we have a uniform value on the z-axis for each value on the x-axis. That is, given a capture time-interval, the sum of the number of messages is uniform for each captured node. This is explained because, given a capture interval, the detection time is uniform for each node captured. Thus, the node runs the CMC protocol for a uniform time for each node captured. Note that, the INFOCOM graphs (Figures 5.(c) and 5.(d)) show some isolated points. For example, in the simulations where nodes 6 or 15 are captured in the first interval (i.e. after 3 hours from the first event in the trace), the number of messages is higher than the number of messages generated when the nodes are captured at the end of the simulation periods (i.e. after 24 hours from the first event in the trace). This is because the first meetings of nodes 6 or 15, with any other node of the network, occurs after the first 3 hours. If we simulate the capture of node 6 or 15 before they meet any other nodes, then the nodes 6 and 15 are not monitored by any node of the network. As a consequence, the capture of node 6 or 15 is not detected, and the simulation does not stop until the end of the INFOCOM trace. This explains why nodes 6 and 15 are quite apart from the remaining nodes in the graphs.

5 Conclusions

In the first part of the paper, we reviewed and improved our previous solution for the node capture detection that leverages node mobility and node cooperation. Later, we investigated the influence of a realistic mobility scenario over a benchmark mobility model (Random Waypoint Mobility Model), using as underlying protocol the above proposal. We showed via extensive simulations the relevance of the mobility model over the achieved performances, measured in terms of: detection time; number of messages and network flooding; and, number of meetings between the nodes. Results indicate that in mobile ad-hoc networks the quality of the solution provided is satisfactory only when it can be adapted to the nodes underlying mobility model.

Our future work aims at investigating detection protocols leveraging: i) the specific mobility pattern the detection protocol is applied to; and, ii) topology maintenance protocol— [9].

References

1. Bandyopadhyay, S., Coyle, E.J., Falck, T.: Stochastic properties of mobility models in mobile ad hoc networks. IEEE Transactions on Mobile Computing 6(11), 1218–1229 (2007)
2. Bettstetter, C.: Topology properties of ad hoc networks with random waypoint mobility. SIG-MOBILE Mob. Comput. Commun. Rev. 7(3), 50–52 (2003)
3. Capkun, S., Hubaux, J.-P., Buttyán, L.: Mobility helps security in ad hoc networks. In: MobiHoc 2003 (2003)
4. Conti, M., Di Pietro, R., Mancini, L.V., Mei, A.: A randomized, efficient, and distributed protocol for the detection of node replication attacks in wireless sensor networks. In: MobiHoc 2007 (2007)
5. Conti, M., Di Pietro, R., Mancini, L.V., Mei, A.: Emergent properties: Detection of the node-capture attack in mobile wireless sensor networks. In: WiSec 2008, pp. 214–219 (2008)
6. Curtmola, R., Kamara, S.: A mechanism for communication-efficient broadcast encryption over wireless ad hoc networks. Electronic Notes in Theoretical Computer Science (ENTCS) 171(1), 57–69 (2007)
7. Demirbas, M., Song, Y.: An RSSI-based scheme for sybil attack detection in wireless sensor networks. In: WOWMOM 2006 (2006)
8. Di Pietro, R., Mancini, L.V., Mei, A.: Energy efficient node-to-node authentication and communication confidentiality in wireless sensor networks. Wireless Networks 12(6), 709–721 (2006)
9. Gabrielli, A., Mancini, L.V., Setia, S., Jajodia, S.: Securing topology maintenance protocols for sensor networks: Attacks and countermeasures. In: SecureComm 2005, pp. 101–112 (2005)
10. Grossglauser, M., Vetterli, M.: Locating nodes with EASE: last encounter routing in ad hoc networks through mobility diffusion. In: INFOCOM 2003 (2003)
11. Hayashibara, N., Cherif, A., Katayama, T.: Failure detectors for large-scale distributed systems. In: SRDS 2002 (2002)
12. Hyytiä, E., Lassila, P., Virtamo, J.: Spatial node distribution of the random waypoint mobility model with applications. IEEE Transactions on Mobile Computing 5(6), 680–694 (2006)
13. Information Processing Technology Office (IPTO) Defense Advanced Research Projects Agency (DARPA). BAA 07-46 LANdroids Broad Agency Announcement (2007), http://www.darpa.mil/IPTO/solicit/open/BAA-07-46_PIP.pdf

14. Liu, H., Wan, P.-J., Liu, X., Yao, F.: A distributed and efficient flooding scheme using 1-hop information in mobile ad hoc networks. IEEE Transactions on Parallel and Distributed Systems 18(5), 658–671 (2007)
15. Luo, J., Hubaux, J.-P.: Joint mobility and routing for lifetime elongation in wireless sensor networks. In: INFOCOM 2005 (2005)
16. Mei, A., Stefa, J.: SWIM: A simple model to generate small mobile worlds. In: INFOCOM 2009 (2009)
17. Newsome, J., Shi, E., Song, D., Perrig, A.: The sybil attack in sensor networks: analysis & defenses. In: IPSN 2004 (2004)
18. Perrig, A., Stankovic, J., Wagner, D.: Security in wireless sensor networks. Commununications of ACM 47(6), 53–57 (2004)
19. Piro, C., Shields, C., Levine, B.N.: Detecting the sybil attack in mobile ad hoc networks. In: SecureComm 2006 (2006)
20. Scott, J., Gass, R., Crowcroft, J., Hui, P., Diot, C., Chaintreau, A.: Crawdad data set cambridge/haggle (v. 2006-01-31) (January 2006),
    `http://crawdad.cs.dartmouth.edu/cambridge/haggle/imote/infocom`
21. Sharma, G., Mazumdar, R., Shroff, N.B.: Delay and capacity trade-offs in mobile ad hoc networks: A global perspective. In: INFOCOM 2006 (2006)
22. Striki, M., Baras, J., Manousakis, K.: A robust, distributed TGDH-based scheme for secure group communications in MANET. In: Proceedings of ICC 2004 (2006)
23. Wang, L., Olariu, S.: A two-zone hybrid routing protocol for mobile ad hoc networks. IEEE Trans. Parallel Distrib. Syst. 15(12), 1105–1116 (2004)
24. Yoon, J., Liu, M., Noble, B.: Random waypoint considered harmful. In: INFOCOM 2003 (2003)

RDTN: An Agile DTN Research Platform and Bundle Protocol Agent

Janico Greifenberg[1] and Dirk Kutscher[2]

[1] Dampsoft GmbH
Vogelsang 1
24351 Damp
Germany
jgre@jgre.org

[2] NEC Laboratories Europe, Network Research Division
NEC Europe Ltd.
Kurfürsten-Anlage 36
69115 Heidelberg
Germany
dirk.kutscher@nw.neclab.eu

Abstract. This paper describes a new approach to developing and evaluating protocols and applications for Delay-Tolerant-Networking based on RDTN, a DTN bundle protocol agent (BPA) implementation written in Ruby. RDTN is light-weight and flexible so that it can be used in DTN application and protocol development as well as in research of DTN-related topics such as routing or convergence layers. RDTN provides a programming language-independent interface for client applications, an interactive environment for tests and a simulation mode for running multiple instances of the RDTN code in a simulated network environment. This paper describes the design of RDTN and its application to agile DTN protocol development.

Keywords: DTN, Implementation, Simulation.

1 Introduction

Delay-Tolerant Networking [1] has evolved from a pioneering research platform towards a well-defined, experimental protocol suite with complete specifications of the overall architecture [2], of the bundle protocol [3], and of different convergence layer protocols. These specifications have been proven to be implementable by the DTN Reference Implementation [4] and other implementations. The main DTN characteristics are: asynchronous, message-oriented store-carry-and-forward-based communications, often leveraging opportunistic networking, with a late-binding-based addressing framework.

The focus of the current research and development work has broadened from investigating the general properties of message-oriented store-carry-and-forward communications towards other topics, including but not limited to: DTN routing [5], [6], [7], applying DTN to various scenarios [8], [9], [10], convergence layers [11], [12], DTN-based overlay protocols [13], [14] and various DTN-based applications [15].

H. van den Berg et al. (Eds.): WWIC 2009, LNCS 5546, pp. 97–108, 2009.

Research and development work in these areas often requires development of new complementary protocols and applications that need to be validated in different scenarios. In the early stages of the development of new protocols, it is particularly important to be able to quickly assess the performance and to check the correctness of the implementation. Based on these test, the protocol can be iteratively improved, before much effort is wasted on sub-optimal first attempts. In allusion to software engineering terminology, we call this approach *agile protocol development*. The quick feedback loop required for this approach is based simulations, as they allow for analyzing the behaviour of networks in larger scenarios than what is achievable with experimentation alone. Simulations can also provide deterministic and reproducible conditions.

RDTN was designed as a platform for DTN research and agile DTN protocol development. It is a simple implementation of the bundle protocol [3] and can be easily and quickly extended for DTN-related research experiments, e.g., for developing convergence layers, routing protocols, and applications. RDTN is usable, robust, and portable, so that it can be used in field tests of delay-tolerant networks. It also includes a simulator, which enables developers to evaluate their DTN development code without any modification in a controllable simulator environment – providing typical advantages of simulated execution: automated simulations, reproducable conditions, independence from real-time clocks etc. RDTN is implemented in Ruby[1] – a dynamic programming language enabling simple and concise solutions to a wide range of programming problems. E.g., as a dynamic programming language, Ruby enables us to dynamically extend the DTN router by adding new protocol functions without requiring explicit compiling and linking steps, which also fits well to the late-binding addressing concepts of the DTN architecture, because code can be loaded into a running process when incoming data requires special handling not included in the basic functionality.

This paper is structured as follows: Section 2 compares RDTN with other Bundle Protocol implementations. Section 3 presents RDTN's main architectural characteristics and describes its components. Section 4 explains the discrete event simulator integrated in RDTN. Section 5 concludes this paper and highlights directions for future work.

2   Related Work

The architecture [2] developed by the DTNRG[2] introduces a *bundle layer* as an overlay for interconnecting different challenged networks. Nodes using this overlay communicate by sending asynchronous messages called *bundles*. The overlay can be attached anywhere in the stack, with *convergence layer adapters* providing an interface to the lower layers. The DTN architecture does not assume the existence of an end-to-end path between senders and receivers, so that bundles are delivered using the store-carry-and-forward paradigm. Moreover, the architecture does not require a single addressing structure to be used throughout the network. Instead, *late-binding* is performed on the endpoint identifiers (EID) which means that nodes only bind the EID to the parameters of a convergence layer when it becomes necessary.

---

[1] http://www.ruby-lang.org
[2] http://www.dtnrg.org

Currently, there are several implementations of the DTN architecture and the bundle protocol. DTN2 [4] is the reference implementation by the DTN Research Group and is considered the most complete implementation of the DTN specifications including proposed extensions. DTN2 can be extended using IPC-protocols for external routing, storage, and convergence layer modules. However, these mechanisms are restricted in their flexibility as they need to adhere to a strict protocol definition. There is also a mechanism to compile DTN2 into a discrete event simulator. RDTN is interoperable with DTN2 in bundle processing functionality and the UDP and TCP convergence layers.

Interplanetary Overlay Network (ION)[3] is an imlementation of the DTN bundle protocol focused on networks between spacecraft and other systems involved in space flight missions. ION is interoperable the with reference implementation, and it includes a simulator for links affected by delays imposed by signal propagation between nodes. However, ION does not allow for simulations of terrestrial mobile networks.

IBR-DTN [16] and DTNlite [17] are implementations optimized for small devices. DASM[4] is an implementation for the Symbian mobile phone operating system. While these implementations are well-suited to real-world testbeds, they are not intended to be used for rapid development and testing of new approaches.

PyDTN[5] is another implementation of the bundle protocol in a dynamic language – Python[6] in this case. It is interoperable with the DTN2 reference implementation, but currently still at a very early development stage. While PyDTN and RDTN have the dynamic implementation language in common, RDTN has a greater focus on being a research platform by including a simulator.

The Opportunistic Network Environment (ONE) simulator [18] is a modular tool that includes a mobility generator, a DTN simulator, and a visualization front-end. ONE implements a variety of DTN routing protocols that can be simulated in scenarios either imported from external traces or synthetically generated. While simulations in ONE allow for conclusions about the simulated algorithms, implementations for actual deployments need to be evaluated separately, as the implementation needs to be specifically tailored to ONE's API. RDTN, on the other hand, can be used both for simulations and deployments, thus allowing for conclusions to be drawn about algorithms and implementations at the same time.

3   Architecture

A key concept of the RDTN bundle protocol agent is the notion of a *flexible, modular and extensible set of building blocks* that are connected through an *event-based coordination mechanisms*. This sections describes the motivation and the novel features of this approach with respect to DTN development.

An instance of RDTN together with the applications using it, constitutes a *node*. A node is structured into conceptual *components*: convergence layers, bundle handling, contact management, routing, and persistent storage. Although the components need

---

[3] https://ion.ocp.ohiou.edu/

[4] http://www.netlab.tkk.fi/~jo/dtn/index.html#dasm-code

[5] http://dtn.sourceforge.net/hg/pydtn

[6] http://www.python.org/

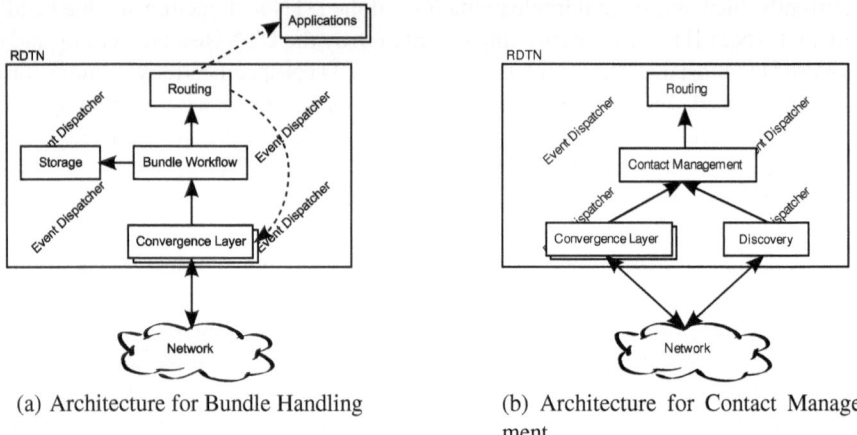

(a) Architecture for Bundle Handling

(b) Architecture for Contact Management

Fig. 1. RDTN Architecture

to communicate with each other and share state, they are loosely coupled. When the state of a component changes in a way that is relevant to other components (e.g., a convergence layer detects that the connection to another node is broken), it creates an event that is broadcast to all components interested in this incident (e.g., a lost link is relevant for the contact management and routing components).

Figure 1 shows two perspectives on the RDTN architecture based on the most important processes. Figure 1(a) illustrates the first top-level process for RDTN: bundle handling. Incoming data from the network is received by a *convergence layer* component extracting the bundle data and passing it on over the event dispatcher. RDTN includes convergence layers for TCP [19], UDP, and FLUTE [20]. Next, the *bundle workflow* component runs the data through a parser, lets the *persistent storage* save a copy of the bundle and handles administrative records and custody transfer. After that, the *routing* component gets the bundle so that it can decide whether to pass the bundle to an application or if it should be forwarded to another node. In the latter case, the bundle is passed to a convergence layer (not necessarily the one over which is was received) for transmitting it over the network.

When a local application is registered to receive a bundle, the routing component forwards it over a special convergence layer that serves as *application interface*. The application interface uses a language-independent protocol based on TCP, so that it can be used over a local network (e.g. in vehicles). In addition to sending and receiving bundles, the interface allows applications to manage registrations and to create acknowledgements.

Figure 1(b) shows how RDTN manages contacts to neighboring DTN nodes. New contacts can be detected either directly by the convergence layers or be the *neighbor discovery* component. A convergence layer detects a contact when another node initiates the communication by sending data or by establishing a connection. Neighbor discovery actively searches the current network environment for other nodes using IP multicast beacons or mDNS messages.

When a convergence layer or the discovery component detect a new node, they generate an event handled by the *contact manager*. The contact manager maintains a list of all available contacts. The information about the new contact is passed to the routing component so that it can consider the new node as receiver of bundles.

To facilitate the loose coupling in RDTN, events are distributed between the components by an event dispatcher. An RDTN node has one instance of this class and all components have a reference to it. The event dispatcher implements a simple mechanism for sending named events and subscribing to them. This event mechanism also allows extensions to be notified of the state of any component without the component being aware of the extension. An event is identified by a symbolic name. The event dispatcher uses this identifier to map an event to all subscribing components. Events can have arbitrary arguments, which the dispatcher passes indiscriminately from source to sink.

When a component subscribes to an event, it passes the event identifier and a Ruby block – i.e. a closure – to the event dispatcher. Each time the event is triggered, the event dispatcher, executes the block, passing all the event's parameters.

When we look at a simple example with a sender and a receiver class, which pass an event with a string parameter between them, the receiver subscribes to the event like this:

```
class Receiver
  def initialize(evDis)
    evDis.subscribe(:testEvent) do |str|
      puts "TestEvent received: #{str}"
    end
  end
end
```

Receiver objects get the event dispatcher in the constructor as is the case for most RDTN classes. It subscribes :testEvent and prints a message including the event's parameter on receiving an event of that type.

To dispatch the event, a sender that dispatches the test event when a new instance is created can be implemented as follows:

```
class Sender
  def initialize(evDis)
    evDis.dispatch(:testEvent, "An event occured")
  end
end
```

RDTN uses multi-threading for managing parallel IO- and timer-related tasks with blocking potential. Although multi-threading comes with the risk of concurrency problems, we decided against using the select-call for IO multiplexing for portability reasons. Neither did we want to use a higher level multiplexing mechanism to avoid having any dependencies besides the Ruby standard library. The threaded approach allows implementers of extensions to use any scheme for multiplexing they deem appropriate as long as it is thread-safe. A new thread is started for each task that may incur long waiting periods. These tasks include waiting for incoming connections and data in convergence layers, sending data, and time-triggered functions, such as deleting expired bundles.

RDTN facilitates the development and evaluation of experimental routing algorithms by providing a basic forwarding logic which can be extended by routing implementations. The basic forwarder provides a simple duplicate detection, interacts with convergence layers, and handles errors. For duplicate detection, the forwarder does not send a bundle over a given link, if the neighbor node that is the remote end of that link has already seen the bundle. Whether a neighbor has seen a bundle is determined by inspecting the log associated with the bundle to see if the bundle was received from or forwarded to the neighbor before. As this mechanism takes only local state into account, it cannot detect routing loops involving more than two peers. More general loop detection must be provided by the concrete routing schemes, if it is required.

The current version of RDTN includes three concrete routing schemes: a static table-based router, epidemic routing [21] and an implementation of the DTN Publish/Subscribe protocol (DPSP) [13].

4   Simulations

For *agile protocol development* for DTN, we want to be able to quickly implement and test new ideas – and then refine the ideas and the implementation and reiterate. As physical deployments incur to much overhead for this iterative development process, we would rather resort to simulations – that can run faster than real-time and allow for an immediate reaction to observations. Furthermore, we want to be able to specify networking scenarios and the interactions between nodes in a simple manner, also allowing for automatic execution with varying parameters.

RDTN includes a module for discrete event simulation that allows us to evaluate the performance of distributed algorithms (e.g., routing schemes or DTN custody transfer) and to test their implementation. It also contains an extension to Ruby's unit test framework that allows researchers to write test cases that are validated against simulation results. Furthermore, RDTN includes helper applications for running sets of simulations with varying parameters.

The RDTN simulator runs multiple nodes, each of which is an instance of the RDTN code. All nodes in a simulation run in one Ruby interpreter, but they are still independent, because each uses its own configuration and event dispatcher. All communication between the simulated nodes is handled by convergence layers as if they were "real" nodes connected over a network. The implementation of simulated nodes is directly derived from the actual daemon and shares its code. Because of this derivation, the node automatically provides the main components of an RDTN instance: persistent storage, routing algorithms, and contact management. The simulation convergence layer passes bundles from one node to another by copying them in memory. This allows us to control the transmission behaviour (e.g., applying rate limits).

Although the simulation could use a convergence layer that also works over a network such as the TCP or UDP convergence layers, we implemented a simulation-specific convergence layer that passes bundles from one node to another by copying them in memory. This allows us to control the transmission behaviour (e.g. applying

rate limits) and we can avoid having to find available ports on the simulation host machine. Synchronization between the nodes also becomes easier to implement.

The simulator uses a simplified model of the links it uses: Links are either connected
or disconnected, have a fixed transmission rate, and no packet losses. We do not try to
simulate the behaviour of concrete convergence layers, as we are mostly interested in
the effects of routing decisions on the network. The limited model of the links reflects
the limited data available from traces of mobile networks, where only connections and
disconnections are logged.

Nodes and the links between them are managed by the *simulation core*. The core
together with the nodes and the simulation convergence layer provides the infrastructure
for simulating delay-tolerant networks; in order to run a simulation, a *network model*
and a *workload* are needed.

A network model is a sequence of connection and disconnection events. These events
can originate from different sources such as parsers for mobility traces or from graph
objects defining networks. The RDTN simulation module contains parsers for mobility
traces in the format used as input for the ns2 network simulator[7], for connectivity reports generated by the dtnsim2 simulator, and for traces from the DieselNet testbed[8]. It
is also possible to use mobility patters processed by the Opportunistic Network Environment (ONE) [18] which includes the Working Day Movement Model (WDM) [22].

A workload is a model for the data to be sent between the nodes so that simulations
generate relevant results about the behaviour of a bundle protocol agent or applications
using it. Workloads for the RDTN simulator are implemented in Ruby using the simulator core to access the nodes and the event queue. The workload can either schedule
actions such as sending bundles by adding events or it can connect to events that are
triggered inside a node. E.g., a workload could be implemented to send a response
whenever a certain node receives a bundle.

When the simulation is finished, statistics need to be gathered, so that the results can
be analyzed. RDTN encodes the statistic data in two structures: the network model that
stores the time-dependent connectivity graph, and the traffic model where the data that
was transmitted between the nodes saved.

The network model can be queried for the properties of the simulated network such
as the number of contacts, the duration of contacts, or the degree of nodes.

The traffic model gives access to delivery statistics, such as the number of sent and
received bundles the delivery ratio, failed transmissions, bundles sizes, etc.

In the following sections, we present two ways in which RDTN simulations can be
used to facilitate research: (1) to define test cases for newly developed protocols, and
(2) to run sets of simulations with varying parameters for exploring the effect of the
parameters and the performance of different proposals.

4.1 Simulation-Based Test Cases

The following example is a unit test that uses a simulated network with two nodes.
One node sends a bundle to the other – which then sends a bundle in return. With the

---

[7] http://www.isi.edu/nsnam/ns/

[8] http://traces.cs.umass.edu/index.php/Network/Network

test framework, we verify that the return bundle was actually received. The following exemplary test case asserts that all bundles are delivered in a simulation with a simple two-node network, where one node sends bundles to the other.

```
01 class TestStress < Test::Unit::TestCase
02    simulation_context 'Under stress the simulator' do
03       network  :two_connected_nodes
04       workload :sending_many_bundles
05       should 'deliver all bundles' do
06          assert_equal 1, traffic_model.deliveryRatio
07       end
08       should 'have two nodes' do
09          assert_equal 2, network_model.numberOfNodes
10       end
11    end
12 end
```

RDTN's test framework is based on *Shoulda*[9], an extension to Ruby's built-in unit test framework. Tests are structured into *simulation contexts* (line 2), that specify the parameters of a run of the simulator and tests validating the results. The input parameters consist of a network model (line 3) and a workload (line 4) which are defined in an external file.

The assertions (lines 5 - 10) can access the results through a *network model* that encapsulates the connectivity graph and a *traffic model* that contains information about the bundles sent over the network. Like the simulation context, the assertions are identified by brief descriptions (lines 6 and 9). The assertions themselves (lines 6 and 9) are implemented in a standard unit test fashion.

4.2   Simulation Specs

While the test cases are ideal for validating basic protocol functionality in deterministic settings, we also need to run exploratory simulations using larger scenarios and sets of simulation parameters, e.g., different routing algorithms or different buffer sizes. The individual simulation runs can take some time, so starting them manually one after another should be automated. Our RDTN-based approach is to create specifications for the simulations – written in a Ruby-based Domain Specific Language (DSL); one for a set of experiments. The variations are listed in the specification, but a run script takes care of computing all combinations for the subsequent runs.

As an example, we test the behavior of a network when sending bundles of different sizes at different rates. More specifically, let us say we want to vary payload size from 1K to 1000K and the sending rate from 1 bundle to 10 bundles per hour in order to analyze the effect on the overall network behavior – simulating all 4 possible combinations. Our simulation spec looks like this:

---

[9] http://dev.thoughtbot.com/shoulda/

```
01 class Example < Sim::Specification
02   def execute(sim)
03     g = Sim::Graph.new
04     g.edge 1 => 2
05     sim.events = g.events
06     sim.nodes.router :epidemic
07     data = 'a' * variants(:size, 1024, 1024000)
08     sim.at(variants(:sendRate, 3600, 360)) do |time|
09       sim.node(1).sendDataTo(data, 'dtn://kasuari2/')
10       time < 3600*24
11     end
12   end
13 end
```

For the sake of this illustration we use a simple network with two permanently connected nodes (lines 3 - 5) using epidemic routing (line 6). In line 7 we declare the first variable parameter: the payload size. By calling Specification#variants, we define a list of variants with the identifier :size. When we run the simulations, this function will either return the first or the second value varying between the runs. We use this size value to create a string of that length to define the actual payload data. We define a time-dependent action that is executed either after one hour (3600 seconds) or six minutes (360 seconds) depending on which variant is currently being executed (line 8). We let node 1 send a bundle with the payload data we assigned in line 7 to node 2 (line 9). When the block executed by the at method returns true, it will be called again after the same amount of time. In line 10, we define that we want the simulator to stop after one day.

To compute the combinations of variants, RDTN needs to go through the spec before starting the actual simulations in a *dry run*. The dry run calls the execute method, passing in an empty Sim::Core object. The difference between a dry run and the actual execution, is the behavior of the variants method. In dry run mode, variants takes the parameter list which contains all the variants and stores it in a hash that serves as a template for generating the combinations. The variants are indexed by the id that is passed as the first parameter to variants (:size and :sendRate in our example).

The template hash of our example looks like this:

```
{:size => [1024, 1024000], :sendRate => [3600, 360]}
```

After the dry run, we compute a list of hashes in which each variant id is mapped to exactly one variant from the template hash that we populated in the dry run. Together, all the hashes in the list cover all combinations of variants. In our example this list contains the following entries:

```
{:size => 1024,    :sendRate => 3600}
{:size => 1024,    :sendRate => 360}
{:size => 1024000, :sendRate => 3600}
{:size => 1024000, :sendRate => 360}
```

For each combination, the run script creates one instance of the simulator core that is passed to the execute method call for the spec. When those executions call the variants method, the value from the current hash is returned. So, when running the first variant, the variant method for `:sendRate` returns 3600, in the second run, we get 360 and so on.

These simulation runs can be executed completely independent from each other, so that they can run in parallel. This allows us to save time making use of multi-core CPUs even if the individual simulation processes are single-threaded. Furthermore, it is possible to run the different variants on different machines. RDTN provides scripts that allow simulations to be executed on Amazon's EC2[10] service that allows developers to dynamically rent computing resources. The variants of the simulation are distributed between the EC2 instances with the help of the Simple Queue Service (SQS)[11]. Each instance automatically gets the next variant from the queue, runs the corresponding simulation and uploads the results. This process continues until the queue is empty and all variants have been simulated.

5   Conclusions

In this paper, we have presented RDTN describing its design and how it can be used as an (agile) research platform. The flexible, light-weight architecture and the use of a dynamic programming language allow for rapid development for extensions implementing proposed research protocols. The integrated simulator allows researchers to test ideas quickly and adapt them accordingly. It is even possible to use a testing framework based on Ruby's unit tests, to continually verify the functionality of the extensions.

As a research platform, RDTN's extensibility is at least as important as the functionality currently integrated. The loosely coupled design and the dynamic implementation language facilitate fast development of additional functionality. Extensions do not have to be specially compiled and loaded into RDTN as would be the case with many other languages. Instead, they simply get loaded like any other code file.

Useful though simulations are, they are not a panacea for determining the viability of network protocols. As we mentioned in section 4, RDTN uses an extremely simplified model of links, so that it's simulation mode is not suitable for work on Convergence Layers. Where the exact properties of the communication media is essential, RDTN needs to be deployed on real nodes.

As future work, we are planning to work on running RDTN on mobile devices such as mobile phones and embedded routers. Porting RDTN to these devices requires a suitable Ruby interpreter. As many mobile phone operating systems provide a Java Virtual Machine, JRuby[12] an implementation of the Ruby lagnuage in Java is a likely candidate for this project.

RDTN is under active development and has been used for evaluations in research papers. In [11] we evaluated the effectiveness of a unidirectional convergence layer implemented for RDTN, and in [13] we used RDTN to implement and simulate our

---

[10] http://aws.amazon.com/ec2/

[11] http://aws.amazon.com/sqs/

[12] http://www.jruby.org/

DTN Publish/Subscribe routing scheme. RDTN is released under the GNU General Public License (GPL) and is available at https://github.com/jgre/rdtn/.

References

1. Fall, K.: A Delay-Tolerant Network Architecture for Challenged Internets. In: Proceedings of ACM SIGCOMM 2003, Computer Communications Review, vol. 33(4) (August 2003)
2. Cerf, V., Burleigh, S., Hooke, A., Torgerson, L., Durst, R., Scott, K., Fall, K., Weiss, H.: Delay-Tolerant Network Architecture. RFC4838 (April 2007)
3. Scott, K.L., Burleigh, S.C.: Bundle Protocol Specification. RFC5050 (November 2007)
4. Demmer, M., Brewer, E., Fall, K., Jain, S., Ho, M., Patra, R.: Implementing Delay Tolerant Networking. Technical report, IRB-TR-04-020, Intel Corporation (December 2004)
5. Lindgren, A., Doria, A., Schelen, O.: Probabilistic routing in intermittently connected networks. In: The First International Workshop on Service Assurance with Partial and Intermittent Resources (SAPIR) (2004)
6. Balasubramanian, A., Levine, B., Venkataramani, A.: Dtn routing as a resource allocation problem. In: SIGCOMM 2007: Proceedings of the 2007 conference on Applications, technologies, architectures, and protocols for computer communications, pp. 373–384. ACM Press, New York (2007)
7. Spyropoulos, T., Psounis, K., Raghavendra, C.S.: Spray and wait: an efficient routing scheme for intermittently connected mobile networks. In: WDTN 2005: Proceeding of the 2005 ACM SIGCOMM workshop on Delay-tolerant networking, pp. 252–259. ACM, New York (2005)
8. Ott, J., Kutscher, D., Dwertmann, C.: Integrating DTN and MANET Routing. In: ACM SIGCOMM Workshop on Challenged Networks (CHANTS) (September 2006)
9. Hui, P., Chaintreau, A., Scott, J., Gass, R., Crowcroft, J., Diot, C.: Pocket Switched Networks and Human Mobility in Conference Environments. In: ACM SIGCOMM Workshop on Delay-Tolerant Networking (WDTN) (2005)
10. Seth, A., Kroeker, D., Zaharia, M., Guo, S., Keshav, S.: Low-cost Communication for Rural Internet Kiosks Using Mechanical Backhaul. In: MOBICOM 2006 (September 2006)
11. Kutscher, D., Greifenberg, J., Loos, K.: Scalable dtn distribution over uni-directional links. In: NSDR 2007: Proceedings of the, workshop on Networked systems for developing regions, pp. 1–6. ACM, New York (2007)
12. Ramadas, M., Burleigh, S., Farrell, S.: Licklider Transmission Protocol - Specification. RFC5326 (September 2008)
13. Greifenberg, J., Kutscher, D.: Efficient Publish/Subscribe-based Multicast for Opportunistic Networking with Self-Organized Resource Utilization. In: The First IEEE International Workshop on Opportunistic Networking (WON 2008), Ginowan, Okinawa, Japan (March 2008)
14. Karlsson, G., Lenders, V., May, M.: Delay-tolerant broadcasting. In: Proceedings of ACM SIGCOMM Workshop on Challenged Networks (CHANTS), pp. 197–204 (September 2006)
15. Ott, J.: Application protocol design considerations for a mobile internet. In: MobiArch 2006: Proceedings of first ACM/IEEE international workshop on Mobility in the evolving internet architecture, pp. 75–80. ACM, New York (2006)
16. Doering, M., Lahde, S., Morgenroth, J., Wolf, L.: Ibr-dtn: an efficient implementation for embedded systems. In: CHANTS 2008: Proceedings of the third ACM workshop on Challenged networks, pp. 117–120. ACM, New York (2008)
17. Patra, R., Nedevschi, S.: Dtnlite: A reliable data transfer architecture for sensor networks. Technical Report Technical report cs294-1 course project report (2003)

18. Keränen, A., Ott, J.: Increasing Reality for DTN Protocol Simulations. Technical report, Helsinki University of Technology, Networking Laboratory (2007)
19. Demmer, M., Ott, J.: Delay Tolerant Networking TCP Convergence Layer Protocol. Internet Draft draft-demmer-dtnrg-tcp-clayer-00.txt (Work in Progress) (October 2006)
20. Kutscher, D., Loos, K., Greifenberg, J.: Uni-DTN: A DTN Convergence Layer Protocol for Unidirectional Transport. Internet Draft draft-kutscher-dtnrg-uni-clayer-00.txt, IETF (April 2007) (Work in Progress)
21. Vahdat, A., Becker, D.: Epidemic routing for partially connected ad hoc networks. Technical Report CS-200006, Duke University (April 2000)
22. Ekman, F., Keränen, A., Karvo, J., Ott, J.: Working day movement model. In: Mobility-Models 2008: Proceeding of the 1st ACM SIGMOBILE workshop on Mobility models, pp. 33–40. ACM, New York (2008)

Realization Aspects of Multi-Radio Management Based on IEEE 802.21

Christian M. Mueller[1], Harald Eckhardt[2], and Rolf Sigle[2]

[1] Universität Stuttgart
Institute of Communication Networks and Computer Engineering
Pfaffenwaldring 47, 70569 Stuttgart, Germany
christian.mueller@ikr.uni-stuttgart.de
[2] Alcatel-Lucent Bell Labs
Lorenzstr. 10, 70435 Stuttgart, Germany
{harald.eckhardt,rolf.sigle}@alcatel-lucent.de

Abstract. In this article, we present a multi-radio management (MRM) architecture for intelligent access selection and load balancing over multiple radio access technologies. We discuss possible implementations of this MRM architecture and analyze to what extent the IEEE 802.21 'Media Independent Handover' framework can be applied here. Starting from the fundamental building blocks of the multi-radio management architecture, we find several issues with respect to the integration with and the interworking between today's 3GPP and non-3GPP networks. Because support of 802.21 can largely differ from one access technology to another, we propose ways to compensate for these differences and finally present an adapted MRM architecture.

1 Introduction

In a heterogeneous mobile communication network, if more than one radio access technology is available at a given location, an intelligent access selection algorithm is needed to select the currently best-suited access technology for a particular service. In a user-centric approach, this algorithm is located in the terminal and the access selection is under user control. In a network-centric approach, the algorithm is located in some control and management device in the operator's network. This has the advantage that cell load information and network status can be taken into account and thus allows for a more efficient utilization of the available resources. Due to an integration with existing radio resource and mobility management procedures, a network-centric approach enables seamless handovers and allows to maintain a high service quality even in a largely heterogeneous environment.

One of the key issues to a multi-radio management is the inherent cross-layer interaction and interworking problem between largely different radio access technologies (RATs). Multi-radio management components need to know about the currently available access technologies and have to issue commands to lower layers. This is not trivial, given that systems today do not always provide this

H. van den Berg et al. (Eds.): WWIC 2009, LNCS 5546, pp. 109–120, 2009.

information and it is usually not propagated through the protocol stack. In addition, lower layer interfaces are access system-specific, which leads to complex implementations due to the intrinsic heterogeneity of the problem. Over the last years, several research projects have addressed the question how a generic cross-layer interface shall look like, which services it has to provide and how it can best be implemented [1,2]. Standardization bodies have adopted these ideas and now standards emerge that have the potential to largely facilitate the realization of the multi radio management ideas. One of these standards is the recently published IEEE standard 802.21 *Media Independent Handover (MIH)* [3].

The idea of a network-centric multi-radio management with 802.21 has already been addressed by a couple of authors. In [4], the authors suggest to locate an 802.21-enabled multi radio device in the access network of different RATs and focus on vertical handovers and measurement reporting. The authors of [5] assume a single 802.21-enabled resource and mobility manager in the core network and evaluate handover sequence, load balancing algorithms and the resulting additional signaling load on the air interface. Eastwood et. al [6] concentrate on 802.21-based mobility between WLAN and WiMAX networks and provide detailed handover sequences on 802.21 service primitive level.

Although these authors describe how 802.21 can generally be used for multi-radio management, only few attention is given to an integration of an 802.21-enabled solution with existing 3GPP and non-3GPP networks. To fill this gap, we thoroughly investigate a 802.21-based realization of a representative multi-radio management framework, denoted as *MRM*. For each of the building blocks of MRM, we evaluate how 802.21 can be used and how it integrates with different radio access technologies. For GSM, UMTS, WiFi and WiMAX networks, we discuss how the relevant 802.21 service primitives are mapped on RAT-specific functions and which further mechanisms are defined to support MIH services. Finding that support of 802.21 primitives is not the same for all RATs, we propose an adapted implementation of the MRM architecture, which is based on 802.21 where applicable, but also includes other mechanisms where necessary.

The remainder of this article is structured as follows: Section 2 gives details of the network-centric the multi-radio management (MRM) architecture. Section 3 then provides a short introduction to 802.21 and its services. In section 4, we discuss the use of 802.21 services for the main MRM building blocks and analyze to what extent these services are supported by the different underlying radio access technologies. Based on the results from this analysis, in section 5, we describe a modified implementation of MRM. Finally, section 6 concludes our work.

2   Multi-Radio Management (MRM)

The multi-radio management solution considered here provides resource and mobility management over different 3GPP and non-3GPP radio access networks and enables intelligent network-centric access selection and efficient load balancing. It follows the same principle of abstraction as presented in [7] and consists of a technology-specific part and a part containing generalized functions that are the same for all RATs. In the following, we briefly describe the building blocks of MRM. For a more detailed description of MRM, we refer to [8,9]. Please note that these building blocks are common to all network-centric multi-radio management solutions and our further analysis is thus not limited to the MRM architecture.

2.1 Building Blocks and Requirements

Access Selection. This block includes the algorithms used to select one access network for a given service out of a number of available networks, e. g. based on policies or other multi-criteria decision making techniques. (The algorithms will not be further considered here, for details see [8,9].)

Measurement Reporting. This includes configuration and reporting of intra- and inter-RAT link measurements. The network-side component needs to be able to detect when the currently serving system becomes insufficient for an ongoing application. In addition, it has to initiate scan commands for candidate neighbor systems, especially in preparation of an inter-RAT handover.

Handover Execution. This includes the handover negotiation phase and the execution of inter-system handovers, i. e. the change from one access technology to another.

Besides these main functional blocks, there are further issues that are non-trivial to resolve when it comes to an implementation in a heterogeneous environment. A brief description of these issues is given in the following:

Neighbor Information Provisioning. To avoid power-consuming scanning for available RATs, information about neighbor systems has to be provided by the network. To support idle mode terminals in their initial access selection, broadcast or multicast channels are required that are accessible without being registered to the network.

Message Transport. The multi-radio management entities need to exchange signaling messages for measurement configuration and handover execution. The signaling transport channels available in the various RATs differ from each other and the MRM components need to adapt accordingly.

Discovery and Registration. A mobile terminal needs to detect whether a system provides multi-radio management functionality and if so, requires a means to access the network-side management entity.

System Interfaces. A multi-radio management entity needs access to lower layer primitives to interact with an existing resource management. These primitives are system-specific or might even not exist at all, because, at the time a system was designed, it has not been foreseen that they need to be accessed by an external application.

2.2 Functional Architecture of MRM

The MRM architecture consists of three different functional entities as illustrated in Fig. 1. The MRM-TE is located in the user terminal and provides

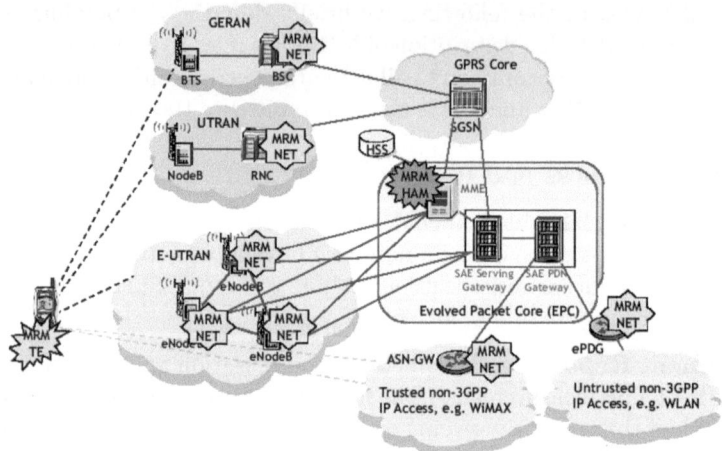

Fig. 1. MRM network architecture

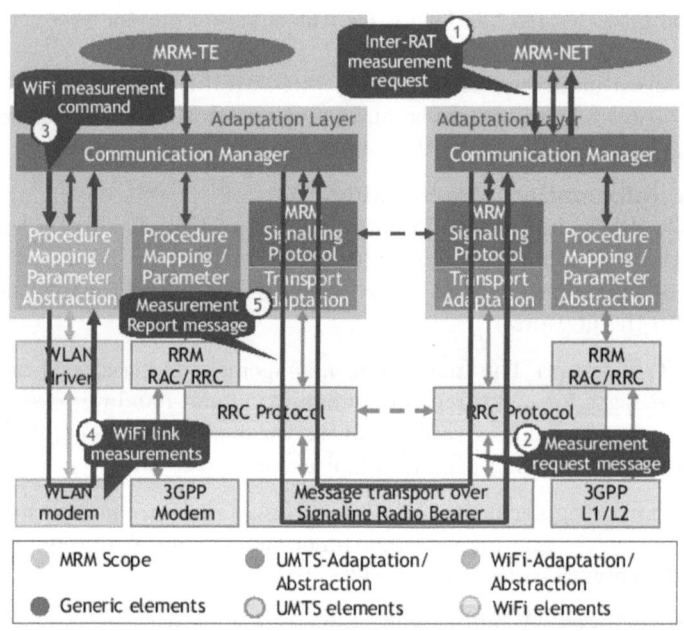

Fig. 2. Inter-RAT measurement procedure for terminals served by UMTS

inter-system measurement functions and an initial access selection algorithm that is used as long as the terminal has not yet established a connection with the access network. The MRM-NET is located in the access network and is associated with all active users within its service area. It communicates with MRM-TE and is implemented on top of the already existing radio resource and mobility management functions of the respective RAT. Its responsibility is to monitor and configure radio and link measurements on user terminals and to trigger inter-system handovers. Therefore, it needs to be involved in existing RRM procedures (e. g. bearer setup, link quality measurements, inter-system handover procedures). Finally, the *heterogeneous access management* element (MRM-HAM) takes access selection decisions based on various input parameters such as link performance, resource usage and availability measurements. The MRM-HAM can either be located as a single MRM server in the core network, or it can be distributed over the respective access network devices together with the MRM-NET.

Figure 2 shows the interaction between MRM-TE and MRM-NET at the example of an inter-RAT measurement request: the terminal is being served by a UMTS network and MRM-NET requests scans for a potentially available WiFi hot spot near its current location (step 1). The measurement request is transmitted over a dedicated signaling radio bearer (2), interpreted by the terminal-side MRM component and a measurement command is issued to the local device driver (3). The measurements are taken (4) and the response is then sent back to MRM-NET(5). Different colors in Fig. 2 indicate which elements are generic, which are part of the underlying RAT and which are needed as an adaptation layer in between to map generic MRM commands to system-specific service primitives.

3   IEEE 802.21 Media Independent Handover

IEEE 802.21 'Media Independent Handover (MIH)' aims at the support of handovers among heterogeneous fixed and mobile networks. 802.21 defines an Information Service (IS) to retrieve and provide mostly static information about current and neighbor networks. The Event Service (ES) provides a standardized set of layer 2 triggers and includes primitives for measurement reporting. The Command Service (CS) defines means to control physical and link layer states and coordinate handover execution. These services are provided by the so-called MIH Function (MIHF) and can be accessed by an MIH User over the MIH Service Access Point, or MIH_SAP. The MIHF does not interpret lower layer triggers or measurements and does not comprise algorithms for access selection or mobility management. This is left to the MIH user. Hence, it merely constitutes an abstraction of RAT-specific lower layer functions, with the MIH_SAP being the uniform interface to access lower layer functions in a media-independent way. 802.21 also includes means to exchange messages among MIHFs and to issue commands to remote MIH-enabled devices by the so-called MIH Protocol, either by using dedicated management frames on layer 2, or by using an IP-based transport protocol.

The IEEE 802.21 has been officially published at beginning of 2009. The analysis here is based on an earlier stable draft version [3]. The draft standard 802.11u 'Enhancements for Interworking with External Networks' [10] defines 802.21-specific extensions for WiFi networks. The recently published amendments 802.11k 'Radio Resource Measurements' [11] and 802.16g 'Management

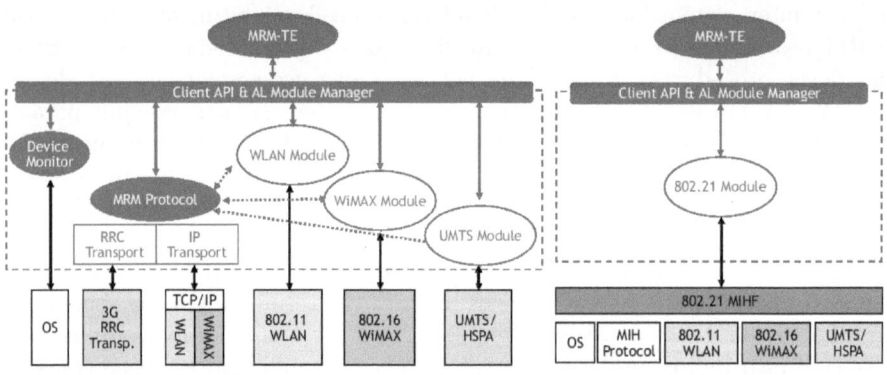

(a) Variant with direct access to RAT primitives   (b) Variant based on 802.21

Fig. 3. Block diagram of MRM entity on the mobile terminal

Plane Procedures and Services' [12] contain further MIH-related functionality for WiFi and WiMAX.

4   Analysis of MRM Based on 802.21

MRM services can be implemented with a RAT-specific adaptation layer for every supported RAT technology. A block diagram of the resulting structure of the MRM-TE is depicted in Fig. 3a. For each RAT (only three different RATs are depicted for simplicity reasons), a dedicated module is required to adapt to the particularities of this RAT. To exchange MRM signaling messages between MRM-TE an MRM-NET, different transport channels have to be used, depending on the current context. To determine the number and type of the available interfaces, an interface to the operating system is required.

It is obvious that such a realization is complex and cumbersome to implement. From the guiding principles of abstraction and generalization [7], the preferable solution would consist of generic MRM components on top of a standardized interface that provides an abstraction from the underlying technology. A standardized management interface then allows to decouple the implementation efforts for the modules above and below this interface and thus facilitates the interworking between and the combination of services from different providers. As an emerging standard, IEEE 802.21 is a strong candidate to provide this interface. A possible realization of the MRM-TE based on 802.21 is depicted in Fig. 3b. In the following, for each of the building blocks identified in section 2.1, we analyze whether the required MRM functionality can be provided using 802.21 primitives. We do this under the specific constraints imposed by current GSM/EDGE and UMTS/HSPA access network architectures, as well as the non-3GPP RATs 802.11 and 802.16. For our analysis, we take 802.21-specific extensions from [10, 11, 12] into account.

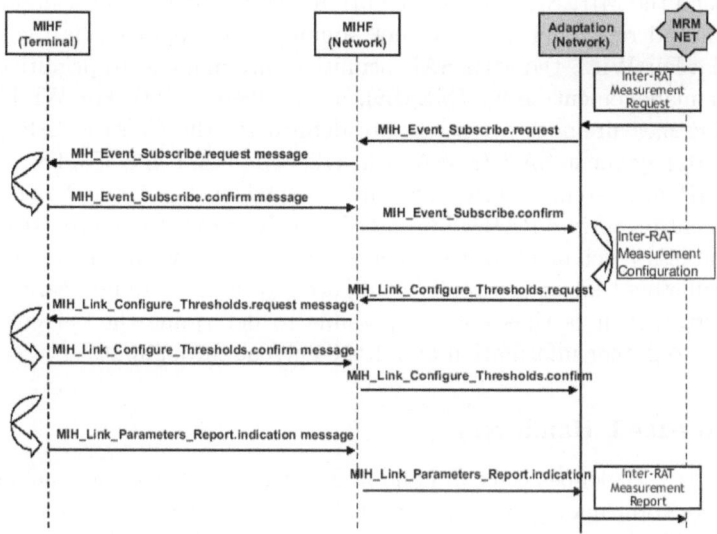

Fig. 4. Measurement reporting signaling diagram

4.1   Access Selection

As it has been mentioned in section 3, 802.21 does not include an access selection algorithm. This is consistent with the MRM architecture described in section 2.2, which sees the access selection as a generic element located above the interface towards the media.

4.2   Measurement Reporting

With the primitives defined for ES and CS, 802.21 provides a flexible mechanism to configure and report measurements of the currently serving and potential neighbor RATs. As an example, Fig. 4 depicts an inter-RAT measurement sequence analog to Fig. 2, now carried out using service primitives of 802.21. It is assumed that the MRM-NET has already discovered which other RATs are available on the mobile terminal, e. g. by using the *Get_Link_Parameters* primitive. The MRM-NET subscribes for measurement reports using the *Event_Subscribe* primitive. The *Link_Configure_Thresholds* command configures the measurements and results are reported in a *Link_Parameters_Report*. In contrast to Fig. 2, there is no MRM adaptation layer on the terminal-side, because the measurement request is handled completely by the MIHF.

The measurement values provided by 802.21 contain several media-independent parameters, e. g. link speed and packet error rate, as well as RAT-specific values, such as the received signal strength indicator RSSI for 802.11. The interpretation of these values is left to the MIH User, respectively the MRM adaptation layer of the access network MRM entity.

Although the MIH_SAP defines the primitives required for measurement con-figuration and reporting, they are not equally well supported by the underly-ing media: For WiFi, the MIH_SAP primitives are mapped to primitives of the sub-layer management entity (MLME), as specified in [11]. For WiMAX, cor-responding measurement primitives are defined for the Control SAP (C-SAP) and the Management SAP (M-SAP) in [12]. For GSM and UMTS, according to [3], MIH measurement primitives are mapped on GSM and UMTS session management primitives RABMSM and SNSM. However, these primitives do not provide access to channel measurements, but only allow to set or retrieve the QoS parameters that are currently configured for a given radio bearer [13]. For 3GPP networks, it is thus not yet possible to determine the current channel quality or to gather information about the current throughput.

4.3   Inter-RAT Handovers

Although 802.21 defines a set of handover primitives, it does not include a mo-bility or location management protocol. Using the MIH handover primitives a mobile terminal can request a handover from a network MIHF entity or the network MIHF can request handover initiation from the mobile terminal. Fur-ther primitives are available to query an MIHF in a target system for avail-able resources and to notify an MIHF when a handover has been completed. In [14], detailed handover sequences between 3GPP and non-3GPP access net-works are already defined. The task of MRM-NET is thus only to trigger these procedures, which can be done using the *MIH_Net_HO_Candidate_Query* and *MIH_Net_HO_Commit* primitives.

4.4   Neighbor Information Provisioning

Information about potentially available neighbor networks and their characteris-tics reduces unnecessary scanning procedures of the mobile terminal. The scanning primitive defined by 802.21 does not provide a way to communicate scanning pa-rameters such as target frequency, scrambling codes or similar. This information thus has to be made available to the terminals using the MIH Information Service.

For WiFi, the *Generic Advertisement Service* [10] allows mobile terminals to access an MIH Information Service database in the network even before an association with the access point has been created. It further allows to send an MIH response as a broadcast to all terminals in the cell.

For WiMAX, with [12] a similar mechanism is in place, by which terminals can send so-called *MIH Initial Service Requests* over the Primary Management Channel before they have completed the network entry procedure. As for WiFi, a query response can be sent as unicast or as broadcast.

For GSM or UMTS, although there are ongoing discussions on a Access Net-work Discovery and Selection Function (ANDSF), which might be based on the 802.21 IS [15], no such functionality has been defined yet. Neighbor information would have to be sent together with other cell list advertisements in system information objects that are sent over the existing broadcast channels.

4.5   Message Transport

For 802.21 services, communication between MIHF entities can be done either on layer 2, or using an IP-based transport protocol. For WiFi, dedicated management frames are defined for L2 transport. For WiMAX, new primitives have been defined to transport MIH messages over the Primary Management Channel. In configurations such as in Fig. 1, where a WiFi or WiMAX access network is connected to a 3GPP core network over intermediate gateways, L2 communication is not possible and an IP-based transport solution is required.

For GSM or UMTS networks, the usage of an IP-based transport is hardly applicable, given that no IP connectivity exists between mobile terminals and radio controllers. IP packets carrying MIH messages would have to be sent to the core network and then be routed back to the currently serving radio controller. This is inefficient and currently not foreseen in the 3GPP architecture. Because no transport mechanism for MIH messages on lower layers is defined, exchange of 802.21 messages among MIHF entities located in a 3GPP radio access network and on the terminal is currently not possible. To resolve this issue, we propose to use the UMTS RR Direct Transfer, respectively the GSM Application Information Transfer mechanism to transparently deliver MIH messages between the mobile terminal and an MRM-NET entity located on a radio controller [16,17].

4.6   Discovery and Registration

A mobile terminal needs to deduce if an access network supports 802.21 respectively MRM functions or not. It further has to determine the address of the network-side multi-radio management entity.

For WiFi and WiMAX, an MIH capability flag has been defined that is part of the beacon frame, respectively the WiMAX Downlink Channel Descriptor. The address of the network-side MIH or MRM entity can be provided in configuration options during the network entry procedure, or by using a DHCP or DNS based address discovery procedure defined for MIH by the IETF Mipshop working group [18].

For GSM and UMTS, there is no MIH capability flag defined so far. A DNS-based discovery is possible if IP-based communication is available. Alternatively, new System Information elements could be specified for this purpose.

5   MRM Realization with Partial Use of 802.21

Summarizing the analysis, it can be observed that current WiFi and WiMAX networks (including the respective amendments) provide sufficient support for 802.21 functions. However, in 3GPP access networks, support for 802.21 services is still very limited. A certain lack of functionality or mismatch of services primitives can be observed with respect to measurement reporting and the support of discovery and registration. A provisioning of information about neighbor networks based on the 802.21 Information Service can not be integrated in 3GPP

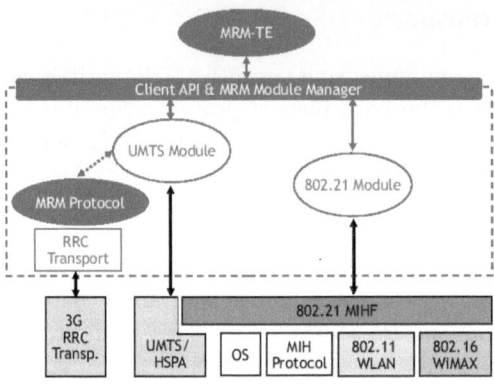

Fig. 5. Partial use of 802.21

Table 1. MRM realization with partial use of 802.21

	Terminal in non-3GPP	Terminal in 3GPP
Measurements (within 3GPP/non-3GPP)	802.21 Command and Event Service	Direct access to 3GPP primitives
Measurements (between 3GPP/non-3GPP)	802.21 Command and Event Service	Indirect measurements via 802.21 and RR Direct Transfer
Neighbor system information	802.21 IS and Generic Advertisement Service	ANDSF (using 802.21 IS) or System Information elements
Inter-System Handover execution	Triggered using 802.21 CS primitives	Direct access to 3GPP primitives
Discovery and registration	DHCP/DNS and MIH capability flag	Dedicated System Information elements
Message transport	IP-based transport of 802.21 messages	RR Direct Transfer

RATs as easily as it is the case for WiFi or WiMAX. The most severe problem arises with respect to the exchange of signaling messages between MIHFs on the terminal side and in the access network, where currently no appropriate transport solution exists.

We therefore conclude that a RAT-agnostic MRM realization on top of 802.21 as shown in Fig. 3b is not yet possible. Assuming that an MRM-NET within a 3GPP access network will not be based on 802.21 for these reasons, we propose a modified implementation of the MRM architecture which distinguishes whether the mobile terminal is currently being served by a 3GPP or a non-3GPP network. Table 1 provides an overview how the different MRM tasks are carried out in these two cases. The resulting structure of the MRM-TE is shown in Fig. 5. While the terminal is being served by a 3GPP network, MRM-NET monitors the current link quality by directly accessing 3GPP measurement primitives, i.e. without using an intermediate 802.21. The same is true for the initiation of handovers. For link availability and link quality measurements of non-3GPP RATs, a 3GPP MRM-NET issues measurement commands to MRM-TE as shown in the right hand side of Fig. 2, using an RR Direct Transfer mechanism. However, on the MRM-TE side, the measurement command is now passed to the local MIHF instead of directly accessing the non-3GPP driver. While the terminal is being served by a non-3GPP network, all procedures are mapped to services provided by 802.21. The advantage of this implementation alternative is that MRM can still make use of 802.21, despite its limited support by 3GPP RATs.

6   Conclusion

For the multi-radio management architecture MRM, we analyzed how its implementation can be facilitated by employing the services provided by IEEE 802.21. We evaluated how these services are supported by an underlying GSM, UMTS, WiFi or WiMAX network. It has been found that, although 802.21 defines a media-independent interface, it is still necessary to distinguish whether a terminal is being served by a 3GPP or non-3GPP access network. This is due to the different degree of support of 802.21 by the respective RATs, especially the absence of signaling transport channels for 802.21 messages in 3GPP access networks. This observation lead to a modified structure of the MRM components that has been presented and discussed here.

References

1. Bandholz, M., Gefflaut, A., Riihijarvi, J., Wellens, M., Mahonen, P.: Unified link-layer API enabling wireless-aware applications. In: IEEE PIMRC (2006)
2. Farnham, T., Gefflaut, A., Ibing, A., Mähönen, P., Melpignano, D., Riihijärvi, J., Sooriyabandara, M.: Toward an open and unified link-layer api. In: Proceedings of the IST Mobile and Wireless Summit 2005, Dresden (Germany) (June 2005)
3. IEEE: IEEE 802.21: Local and metropolitan area networks: Media independent handover services. Draft Standard, v7.0 (July 2007)
4. Lampropoulos, G., Salkintzis, A., Passas, N.: Media independent handover for seamless service provision in heterogeneous networks. IEEE Communications Magazine 46(1) (2008)
5. Melia, T., de la Oliva, A., Vidal, A., Soto, I., Corujo, D., Aguiar, R.: Toward IP converged heterogeneous mobility: A network controlled approach. Computer Networks 51(17) (2007)
6. Eastwood, L., Migaldi, S., Xie, Q., Gupta, V.: Mobility using IEEE 802.21 in a heterogeneous IEEE 802.16/802.11-based, IMT-advanced (4G) network. IEEE Wireless Communications 15(2) (2008)
7. Sachs, J., Aguero, R., Daoud, K., Gebert, J., Koudouridis, G., Meago, F., Prytz, M., Rinta-aho, T., Tang, H.: Generic abstraction of access performance and resources for multi-radio access management. In: IST Mobile and Wireless Communications Summit. (2007)
8. Piao, G., David, K., Karla, I., Sigle, R.: Performance of distributed MXRRM. In: IEEE PIMRC (2006)

9. Blau, I., Wunder, G., Karla, I., Sigle, R.: Cost based Heterogeneous Access Management in multi-service, multi-system scenarios. In: IEEE PIMRC (2007)
10. IEEE: Wireless LAN Medium Access Control (MAC) and Physical Layer (PHY) Specifications, Amendment: Enhancements for Interworking with External Networks. Draft Standard 802.11u (2008)
11. IEEE: Wireless LAN Medium Access Control (MAC) and Physical Layer (PHY) Specifications, Amendment 1: Radio Resource Measurement of Wireless LANs. IEEE standard 802.11k (2008)
12. IEEE: Part 16: Air Interface for Fixed and Mobile Broadband Wireless Access Systems, Amendment 3: Management Plane Procedures and Services. IEEE standard 802.16g (2008)
13. 3GPP: Mobile radio interface signalling layer 3; General Aspects. TS 24.007 (2005)
14. 3GPP: Architecture enhancements for non-3GPP accesses (Release 8). TS 23.402 (2008)
15. 3GPP: System Architecture Evolution (SAE); CT WG1 aspects. TR 24.801 (2008)
16. 3GPP: Radio Resource Control (RRC); Protocol specification. TS 25.331 (2008)
17. 3GPP: Mobile radio interface layer 3 specification; Radio Resource Control (RRC) protocol. TS 44.018 (2007)
18. IETF Mipshop: Mobility Services Framework Design (MSFD). draft-ietf-mipshop-mstp-solution-06 (work in progress) (2008)

Seamless Mobility of Senders Transmitting Multi-user Sessions over Heterogeneous Networks

Luis Veloso[1], Eduardo Cerqueira[1], Paulo Mendes[2], and Edmundo Monteiro[1]

[1] Department of Informatics Engineering, University of Coimbra,
Polo II – Pinhal de Marrocos, 3030-290 Coimbra, Portugal
{lmveloso,ecoelho,edmundo}@dei.uc.pt
[2] Internet Architectures and Networking Telecommunication and Multimedia Unit,
INESC Porto, Rua Dr. Roberto Frias, 378, 4200-465, Porto, Portugal
pmendes@inescporto.pt

Abstract. The development of wireless technologies together with the increasing portability of telecommunication devices lead to the necessity to develop communication systems capable of supporting a seamless mobility experience to end-users. Moreover, the arising of multi-user real-time services has emphasized the importance of the multicast communication to deliver content to multiple simultaneous receivers. The simultaneous increase in the production and distribution of multimedia content by both mobile service providers and consumers requires an efficient solution for the mobility management of multicast sources. This way, mobility management of multicast senders is a challenging and unsolved task for the successful development of next generation user-centric mobile networks, where the nomadic user is not only a consumer but also a provider of multimedia content. A proposal to constraint losses during the handover of multicast senders, and consequently, to avoid the service degradation perceived by receivers is presented in this paper. Simulation results confirm the ability of the proposed mechanism to reduce or completely avoid packet losses, increasing the perceived quality of the received sessions.

Keywords: User-centric Mobile Communications, Multicast, Seamless Multimedia Experience, QoS.

1 Introduction

Wireless networking represents the future in terms of connectivity and ubiquitous access. The current wireless landscape is characterized by distinct radio access technologies developed to fulfill the requirements and expectations of users. From the traditional cellular networks (e.g. GSM, EDGE, UMTS, CDMA2000) to the local area networks (e.g. IEEE 802.11 b/g/n) and wide area networks (e.g. IEEE 802.16 d/e, DVB-T/H, DAB), wireless technologies represent a rapid area of growth and importance. This progress creates a demand for mobility management techniques able to provide a mobile communication without perceived quality degradation in user-centric systems.

H. van den Berg et al. (Eds.): WWIC 2009, LNCS 5546, pp. 121–132, 2009.

The future Internet architecture will also need to support applications aimed to multiple simultaneous users (multi-user), such as mobile IPTV, voice and video conferencing, push media (e.g. news headlines, weather updates), file distribution and monitoring (e.g. sensors, stock). Multicast communication is the most appropriate technique to distribute the content of multi-user real-time services, because it is more efficient than unicast in terms of resource utilization for services shared by groups of receivers.

Since multicast protocols were not developed to support seamless mobility, mobility management is necessary to provide seamless handovers involving this type of communication. For this reason, the *Seamless Mobility of Users for Media Distribution Services* (SEMUD) [1, 2] approach was previously proposed to provide seamless mobility of multicast receivers based on the combination of caching and buffering mechanisms together with mobility prediction and session context transfer among access routers.

Mobility management of multi-user receivers is not a trivial task, and the challenge increases when trying to manage seamless source mobility. This occurs because the movement of a multicast sender has higher impact in the shape and efficiency of the multicast tree than the movement of multicast receivers. The key challenge is that while the receiver movement on a multicast delivery is complex but has a local impact in the multicast distribution tree, the source movement brings more problematic issues since it is global, affecting the complete reconstruction of the multicast tree. The subsequent subscription of a new multicast tree by the receivers will lead to packet losses, and consequently, to the related degradation of session quality.

This work presents an approach named *Seamless Mobility of Senders for Media Distribution Services* (SEMSE) to address the sender mobility challenge. The seamless movement of senders expresses the ability to protect receivers from packet losses which could occur during handover. SEMSE is supported by buffering mechanisms responsible to assist the movement of senders between different access routers in order to guarantee communications without packet losses, taking advantage from the capability of buffers to store packets. The foundation of the proposed mechanism is the combination of a buffer denominated *Holder* and located in mobile nodes with a buffer denominated *Virtual Multicast Root* (VMR) which is placed at the egress edge of the access network. The cooperation between these components will permit to reduce or completely avoid packet losses during the movement of mobile sources. The proposal was evaluated regarding the packet losses avoidance and their influence in the quality of a video session perceived by the final receiver. The *Peak Signal to Noise Ratio* (PSNR) was the metric chosen to measure the quality of the received video sessions.

The referred mechanisms were developed under the *QoS Architecture for Multiuser Mobile Multimedia* (Q3M) architecture [3]. This architecture controls the quality level, connectivity and seamless mobility of multi-user sessions across heterogeneous wired and wireless environments. As a brief overview, the Q3M architecture controls the flow of multi-user sessions across heterogeneous networks through the use of an edge networking approach, in which the functionality of each network is controlled by a group of organized edge devices.

The remainder of this document is organized as follows. Section 2 gives a survey of the related work and section 3 describes the mechanism specification. Next, the simulation scenario and experimental results are discussed in the Section 4. Finally, conclusions are presented in Section 5.

2   Related Work

The most relevant mobility management techniques intended to support the mobility of senders transmitting multi-user sessions are described in this section.

To support multicast transmission over *Mobile IP* [4] networks, the IETF proposed the *Home Subscription*. In this approach, the mobile source uses its *Care of Address* (CoA) to tunnel the multicast data to its home agent. The enclosed data contains the *Home Address* (HoA) as the source address and the multicast group address as the destination address. In the home agent, the data is decapsulated and forwarded to the multicast tree. This approach gives transparency to the handover of a mobile source, since a source-specific tree will be built with reference to the fixed HoA. This is, the multicast tree is rooted in the HoA, and consequently, there is no need to reconstruct the multicast tree whenever a handover of the mobile source occurs. However, this proposal introduces sub-optimal routing and a central point of failure as well as it lacks in seamless support. Moreover, the processing task of the home agent increases with the number of mobile sources, wasting system resources in this entity.

Mobicast [5] proposal introduces a *Domain Foreign Agent* (DFA) aimed to support the mobility of multicast users within a foreign network. In the case of mobile receivers the amount of packet losses is reduced since the multicast tree is subscribed not only by the current base station of the mobile receiver, but also by the adjacent base stations where the arriving packets will be stored. A main drawback of this approach is the inefficient network resources utilization. Furthermore, when the mobile node is sending a multicast session, it encapsulates the multicast packets and sends them to the DFA. The DFA will decapsulate and send them to the multicast tree on behalf of the mobile sender. This way, the reconstruction of the tree is avoided when the mobile source moves but no mechanism is provided to constrain the packet losses occurring during handover. This scheme leaves to the multicast application the course of action regarding the packets lost during handover. On the contrary our proposal places the loss control on the edges of the access network.

Another scheme proposed by the IETF to manage multicast transmission was the *Remote Subscription*, in which the multicast tree is established towards the new location of the mobile source. That is, the mobile source uses its current CoA as the source address of the multicast group. The advantage of this scheme is to provide a shortest multicast forwarding path, thereby avoiding triangle routing across the home network, and to avoid the utilization of tunnels when the mobile node moves between access routers. However, this approach requires the reconstruction of the entire multicast tree when the sender moves, causing service disruption. Two worthy proposals for source mobility exist based on this latter approach: the *Source Mobility support Multicast* [6] (SMM) and the *Mobile SSM Source* [7] (MSSMS).

In the SMM approach, the multicast tree is build over a Cellular IP micro-mobility network, in which a *Shortest Path Tree* and *Rendezvous Point Tree* are combined into one multicast tree. The *Rendezvous Point Tree* is used to minimize the overhead in the tree reconstruction caused by source movement and the *Shortest Path Tree* is used to eliminate redundant routes for the members between the source and rendezvous points. This scheme has the advantage to suppress the overhead of the multicast tree reconstruction. Nonetheless, it depends on the Cellular IP technology and requires enhancements to be applied to other IP networks, not providing seamless mobility control.

The MSSMS describes enhancements to MIPv6 that can be used to solve the problems introduced by mobile *Source Specific Multicast* [8] (SSM) sources. In this scheme, when the mobile source moves into a new network, it notifies the multicast receivers about its *new CoA* (nCoA). Upon reception of this notification, receivers initiate a join operation towards the new channel (nCoA, G) but only prune the old channel when they start receiving packets on the new channel. To ensure consistency at higher layers, the notification must also indicate the HoA of the source to permit the correct identification of the SSM session. At the new foreign network the mobile source sends packets with the source address nCoA to the new multicast tree, and encapsulates the multicast packets to the *old Access Router* (oAR) with the *old CoA* (oCoA) address at the inner header. At a point in time, two multicast trees coexist but for a given receiver, only one branch is active at a time. The source only stops encapsulating the data to the old tree when it is notified by the previous access router that there are no receivers listening to the old channel (oCoA, G). Finally, the new multicast tree with origin at the new CoA of the source is completely formed. With this proposal a smooth transition is accomplished between the two multicast trees. Nevertheless, this approach suffers from redundant routing that causes packet routing delay and network bandwidth consumption during the process of handover.

The analysis of related work reveals that none of the solutions provides an acceptable compromise between seamless mobility and network resource utilization. Therefore, it would be interesting to investigate how the usage of buffers as custodians of multicast packets can provide a good trade-off between limiting the service degradation perceived by the final user and limiting network resource usage.

3 Mechanism Specification

The objective of the *Seamless Mobility of Senders for Media Distribution Services* (SEMSE) approach is to provide seamless mobility to senders of multi-user sessions while limiting the usage of network resources, when senders move between different attachment points. The seamless movement is supported by buffering mechanisms responsible to assist the transmitting mobile nodes on changing the attachment point between different access routers, without or with reduced packet losses. This is achieved through the combination of a buffer in the mobile node referred as *Holder*, and buffers in the access network called *Virtual Multicast Roots* (VMR). Figure 1 presents a generic scenario where the location of the referred components can be identified.

Each access network will contain at least one VMR composed by a cluster of buffers, being each buffer dedicated to a single multicast session. Hereinafter, a single session will be used for illustration purposes, and consequently, the respective VMR will be composed of a single buffer.

The multicast session is sent from the mobile sender to the VMR via a tunnel, where it is stored and sent to the multicast tree. From the view point of the receivers, the multicast source is the VMR being the multicast tree defined by <VMR, G> and not <Sender, G>. This way, the source movement will be concealed from the receivers, which avoids the reconstruction of the entire tree each time the sending mobile source executes a handover. During handover, and despite the lost of connection with the mobile source, the VMR continues sending the buffered packets, avoiding in

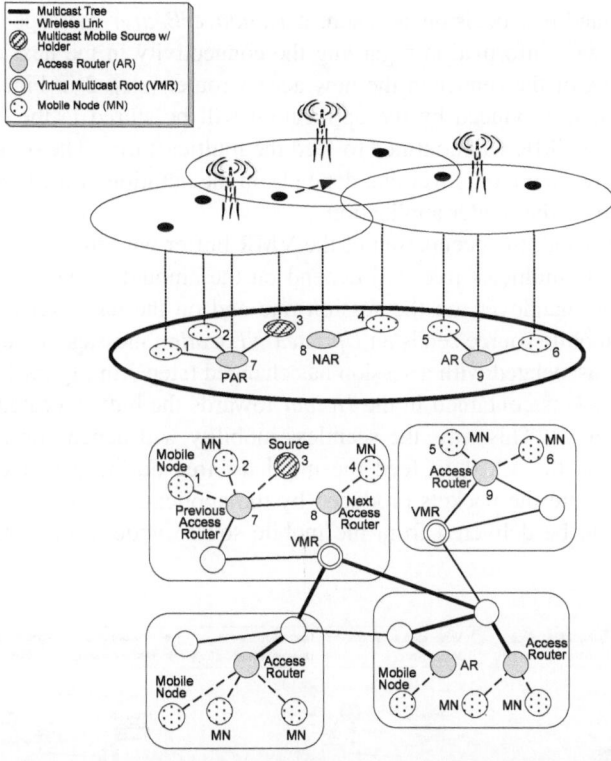

Fig. 1. Scenario describing the location of SEMSE components

this manner the interruption of the session flow. This capability depends on the amount of packets buffered by the VMR, on the transmission bit rate and on the handover duration. The role of the *Holder* is to store the packets produced by the sender application during handover, in order that after this period, its interaction with the VMR will constrain/eliminate packet losses.

When the distance from the mobile source to the VMR becomes excessive, it could be advantageous the joining of a new VMR by the mobile source.

A detailed description of the SEMSE mechanism, along with an example illustrating its functionality, will be presented in the following section.

3.1 SEMSE Operation

As described in Figure 2, the multicast mobile source sends packets towards the VMR via a tunnel (step 1 in Figure 2), from where they are delivered to the multicast tree. This way, the mobile source can send multicast packets even when connected to an access network that does not support IP multicast. It is assumed that each access router has a database with the IP address of the local available VMRs. This database can be created manually or on-demand by performing periodic message exchanges between neighbor access routers as happens in the *Fast Mobile IP* (FMIP) [9] proposal, though the VMRs discovery (and maintain) process is out the scope of this paper.

When the handover decision is taken, a *HandoverBearer* message is sent to the mobile source with information regarding the connectivity in the next access router, namely the CoA of the sender in the new access router (step 2 in Figure 2). During handover, the data produced by the application will be stored in the corresponding *Holder* while the VMR will continue to feed the multicast tree. The occupation of the *Holder* will increase up to a level that depends on the duration of the handover and on the bit rate used by the sender application.

On the other hand, the occupation of the VMR buffer will diminish and the capacity to nourish the multicast tree will depend on the amount of packets stored in the VMR before the handover, on the session rate and on the handover duration. After handover, the mobile source sends an *UpdateMNLocation* message to notify the VMR that its address associated with a session has changed (step 3 in Figure 2). Afterwards, it pushes the packets contained in the *Holder* towards the buffer located in the VMR (step 5 in Figure 2). This way, the seamless mobility will depend on the amount of data available on the VMR to feed the multicast tree during handover and on the *Holder* size to store the packets produced by the application. Subsequently, the session continues to be delivered from the mobile source node to the VMR (step 6 in Figure 2).

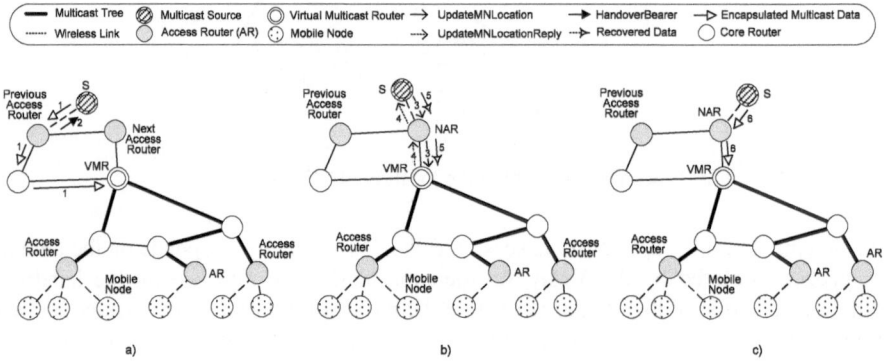

Fig. 2. Operation of the SEMSE mechanism

In this manner, not only the reconstruction of the multicast tree is avoided since the respective source address remains the same from the view point of the receivers, but also packet losses are constrained/eliminated. The SEMSE mechanism is supported by the following messages:

- *UpdateMNLocation*: message sent from the sender mobile node notifying a VMR about its new address;
- *UpdateMNLocationReply*: acknowledge message sent from VMR to the mobile source;
- *HandoverBearer*: message sent by the current access router to the mobile source (after handover decision), providing information related to the future access router and VMR.

3.2 Interfaces and Components

The SEMSE mechanism contains several interfaces and components needed for seam-less source mobility procedures. Firstly, an interface is needed for the interaction with session control mechanisms. When the handover decision is taken, the interaction with session control mechanisms aims to transfer the session context to the new access router, install the session in the new access router and path, as well as to release resources on the old path. Moreover, an internal interface is used to allow the interaction between SEMSE components. For example, when the sender moves a message is sent towards the VMR to update the sender new location.

Holders and VMRs are the components which compose the SEMSE mechanism. As referred before, the first ones are comprised by a buffer used to store packets produced by the sender application whilst a VMR is composed by a cluster of buffers, being each buffer dedicated to a single session.

3.3 Advantages

Between the most relevant benefits of the SEMSE proposal it should be referred its ability to keep media transmission during the handover of a multicast sender, to support micro and macro-mobility and to assure the operation of multicast senders in unicast or multicast domains.

Therefore, it provides transparent support for receivers, which do not perceive any interruption of the flow during the source mobility, by avoiding or significantly reduce packet losses during the handover. This allows network operators to increase the level of satisfaction of users by improving the user experience during handovers.

Moreover, the operation of SEMSE introduces low communication overhead and has low complexity.

4 Evaluation

The discussion that follows tackles the evaluation of the SEMSE mechanism. First, the scenario of the simulations is addressed. Then, the amount of packets that is possible to recover with SEMSE is presented. Afterwards, the impact of the improved packet loss avoidance in the quality reception of a video session is analyzed. Finally, the VMR buffer occupation is captured during the simulation time and presented for a broad understanding of the dynamics involving the system behavior.

4.1 Scenario

The simulation was implemented following the scenario described in Figure 2, where a mobile node is sending a video sequence composed of 300 frames denominated *News*. Those frames were originally in the raw YUV format with 4:2:0 sampling and CIF (352x288) size. They were compressed using a MPEG4 encoder into a video sequence with a *Group of Pictures* (GOP) structure composed by one I-frame and twenty nine P-frames, and transmitted with a 30 frame/s rate. Each frame was fragmented into blocks with 1024 bytes length which were transported in RTP [10] packets. Those packets were sent using a *Variable Bit Rate* (VBR) with an average rate of 86 KB/s.

The intra and inter-network links have a bandwidth of 100 Mbit/s and the wireless link has a capacity of 11 Mbit/s. A handover has been generated at the middle of the simulations being its duration indicated at each of the following descriptions regarding the obtained results. The simulation was implemented using the discrete event simulator NS2 [11] and the open source framework Evalvid [12].

Since in this case a video sequence is considered, the PSNR metric was used to measure the quality of the received session. Considering the luminance (Y) of the received and sent frames and assuming frames with MxN pixels, PSNR is expressed through the following expression:

$$PSNR = 20\log_{10}\left(\frac{255}{\sqrt{\frac{1}{MxN}\sum_{i=0}^{M-1}\sum_{j=0}^{N-1}\|Ys(i,j)-Yd(i,j)\|^2}}\right)$$

In this equation, while Ys(i,j) designates the pixel in the position (i, j) of the original frames, the Yd(i,j) represents the pixel located in the position (i, j) of the received frames. With this metric each received frame will be compared with the original one, given the obtained value a measure for the degradation of the original frame.

4.2 Packet Losses

The quantity of recovered packets versus the VMR buffer size and the handover duration is depicted in Figure 3, where the source buffer and receiver buffer sizes (150 KB) remained constant. As expected, the amount of recovered packets is low and irregular for small VMR buffer sizes (< 28 KB), which are insufficient to accommodate the variable bit rate of the session.

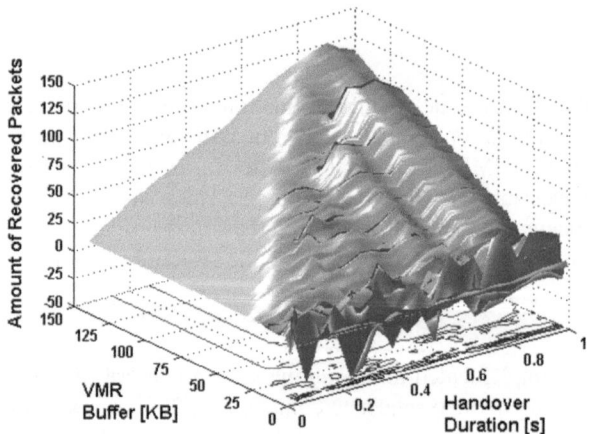

Fig. 3. Recovered packets versus the VMR buffer size and handover duration

For higher values the results show that, considering a constant VMR buffer size, the amount of recovered packets increases with the handover duration and then becomes approximately constant with an oscillating behavior. This oscillation is motivated by the variable bit rate of the transmitted session. This is, for those handover durations which cause the fluctuation, the VMR buffer becomes full with the packets sent after handover by the mobile source buffer. Subsequently, the variable bit rate of the session leads to buffer overflows, thereby causing packet losses whose amount varies as the handover duration increases.

A different behavior occurs when considering a constant handover. The amount of recovered packets increases initially with the VMR buffer size. However, this increase occurs not linearly but with an oscillation due to the variable bit rate of the session, together with a VMR buffer size too small to accommodate the packets sent after handover from the mobile source.

4.3 Peak Signal to Noise Ratio (PSNR)

This section investigates the influence of the proposed mechanism in the quality of a received video session. For this reason the PSNR of the received video sequences was analyzed versus the *Holder* and VMR buffer sizes. PSNR values higher than 37 dB reflect a good quality video, whilst values between 20 and 25 dB indicate a poor video reception.

Figure 4(a) depicts the obtained PSNR versus the *Holder* size when the SEMSE mechanism is disabled. These results were obtained considering a constant VMR buffer size (150 KB), receiver buffer size (150 KB) and handover duration (500 ms). The quality degradation due to packet losses occurring during handover is noticeable.

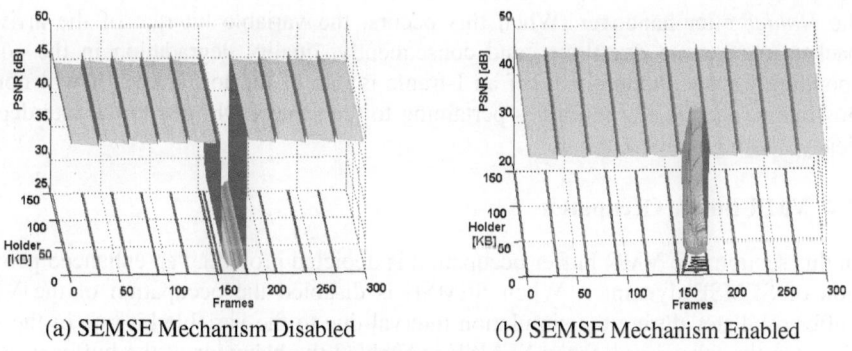

(a) SEMSE Mechanism Disabled   (b) SEMSE Mechanism Enabled

Fig. 4. PSNR versus the Holder buffer size

The improvement introduced by SEMSE is shown in Figure 4(b). In this situation, as the *Holder* size increases more packets are stored during handover and posteriorly sent to the VMR. Consequently, more frames are correctly received and decoded.

The impact of packet losses in the quality of the received session was also evaluated versus the VMR buffer size, considering a constant Holder buffer (150 KB), receiver buffer (150 KB) and handover duration (500 ms). Figure 5(a) depicts the obtained PSNR when SEMSE is disabled. The influence of the handover related losses is perceptible in the PSNR of the frames transmitted during this period.

When SEMSE is enabled, the packets stored during handover in the source buffer are sent to the VMR. As the VMR buffer size increases more packets can be recovered, and consequently, a better PSNR will be expected. Figure 5(b) confirms this supposition since the quality of the previously degraded frames increases when SEMSE is enabled.

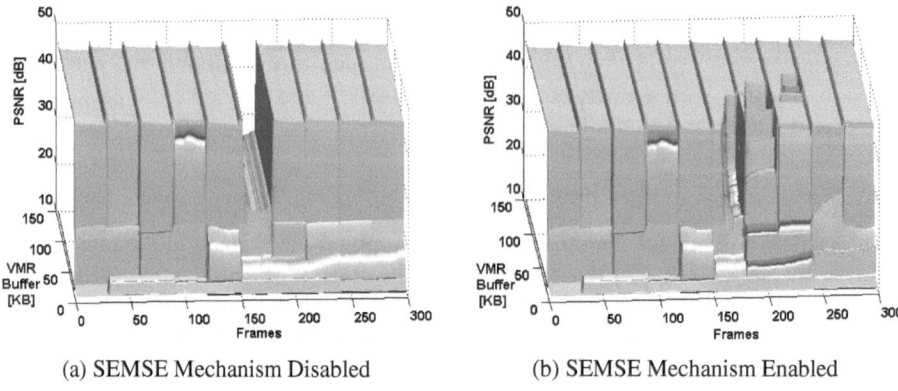

(a) SEMSE Mechanism Disabled   (b) SEMSE Mechanism Enabled

Fig. 5. PSNR versus the VMR buffer size

However, some frames have an unexpected low quality. This occurs because in these cases the VMR buffer size is too small to accommodate all the packets sent by the *Holder* after handover. When this occurs, the variable bit rate of the arriving packets may cause overflows, and consequently, quality degradation in the corresponding frames. Additionally, if an I-frame is one of the lost frames it will not be possible to decode all the frames pertaining to the same GOP, due to the interdependency existing between them.

4.4 VMR Buffer Occupation

In this section, the VMR buffer occupation is depicted providing an enhanced perception of SEMSE dynamics. When SEMSE is disabled the occupation of the VMR buffer oscillates during the simulation interval due to the variable bit rate of the session. On the other hand, when SEMSE is enabled the behavior of the buffer occupation is significantly different as described in Figure 6.

The occupation of the VMR buffer versus the *Holder* was obtained considering a constant handover duration (500 ms) and receiver buffer size (150 KB). The higher peaks result from the fragmentation and transmission of I-frames which are the ones with larger sizes. When the mechanism is enabled, the occupation of the VMR buffer suffers an abrupt increase after the handover procedure, proportional to the *Holder* size. This occurs, because as the *Holder* size increases, more space is available to store packets during handover. Consequently, after handover more packets will be sent from the source to the VMR. Nonetheless, for a constant handover duration this increase is not infinite since the number of packets to recover is finite.

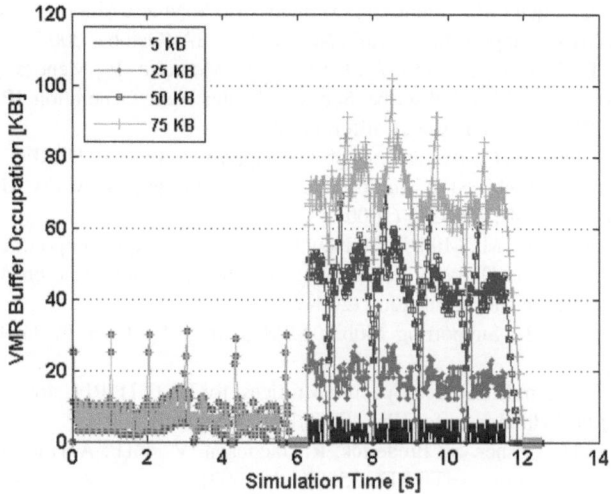

Fig. 6. VMR buffer occupation versus the Holder size

5 Conclusion

This paper presented the SEMSE approach to support the seamless mobility of multicast senders. The obtained results confirm the proposal ability to avoid packet losses during the movement of a mobile source. In addition, it was observed the dramatic impact of the improved packet recovery in the quality of a video sequence. Nevertheless, SEMSE buffers should be correctly dimensioned to absorb rate variations and to efficiently recover the missing packets. For example, for an average rate of 86 KB/s, a handover duration of 500 ms, and source and receiver buffer sizes with 150 KB, the losses will be totally avoided for VMR buffer sizes above 102 KB.

As future work, further evaluation will be done to the mechanism efficiency when the movement of the mobile source leads to a change of the active VMR. Moreover, the optimal number of mobile sources that a VMR is able to cover while supporting seamless mobility will be also investigated. In addition, the proposed mechanism will be analytically modeled to obtain an even better insight of its performance.

Acknowledgments. This work was supported by the NTT DoCoMo Euro-labs, by the Portuguese *Ministry of Science, Technology and High Education* (MCTES), and by European Union FEDER under program POSC (projects Q3M and SAPRA).

References

[1] Veloso, L., Cerqueira, E., Mendes, P., Monteiro, E.: Selective Mobility support of Multi-user Services in Wireless Environments. In: 18th Anual IEEE International Symposium on Personal, Indoor and Mobile Radio Communications (PIMRC 2007), Athens, Greece (September 2007)

[2] Veloso, L., Cerqueira, E., Mendes, P., Monteiro, E.: Seamless Mobility of Users with QoS and Connectivity Support. In: WiMob, New York, USA (October 2007)

[3] Cerqueira, E., Veloso, L., Neto, A., Curado, M., Monteiro, E., Mendes, P.: A Unifying Architecture for Publish-Subscribe Services in the Next Generation IP Networks. In: Globecom 2006, San Francisco, California (2006)

[4] Johnson, D., Perkins, C., Arkko, J.: Mobility Support for IPv6, IETF RFC 3775 (2004)

[5] Tan, C., Pink, S.: Mobicast: a Multicast Scheme for Wireless Networks. Mobile Networks and Applications 5(4), 259–271 (2000)

[6] Sato, K., Katsumoto, M., Miki, T.: A New Multicast Technique Supporting Source Mobility for Future Vision Delivery Service. In: International Conference on Advanced Information Networking and Application (2004)

[7] Jelger, C., Noel, T.: Supporting Mobile SSM Sources for IPv6. In: Globecom, Taiwan (2002)

[8] Holbrook, H., Cain, B.: Source-Specific Multicast for IP, IETF RFC 4607 (2006)

[9] Koodli, R.: Fast Handovers for IPv6, IETF RFC 4068 (July 2005)

[10] Schulzrinne, H., Casner, S., Frederick, R., Jacobson, V.: RTP: A Transport Protocol for Real-Time Applications, IETF RFC 3550 (July 2003)

[11] Ke, C., et al.: An Evaluation Framework for More Realistic Simulations of MPEG Video Transmission. Journal of Information Science and Engineering, 425–440 (March 2008)

[12] Klaue, J., Rathke, B., Wolisz, A.: EvalVid – A framework for video transmission and quality evaluation. In: Kemper, P., Sanders, W.H. (eds.) TOOLS 2003. LNCS, vol. 2794, pp. 255–272. Springer, Heidelberg (2003)

An Adaptive Optimized RTO Algorithm for Multi-homed Wireless Environments

Sheila Fallon[1], Paul Jacob[1], Yuansong Qiao[1], and Liam Murphy[2]

[1] Software Research Centre, Athlone Institute of Technology, Ireland
Sheila.Fallon@gmail.com, {pjacob,ysqiao}@ait.ie
[2] School of Computer Science and Informatics, University College Dublin, Ireland
Liam.Murphy@ucd.ie

Abstract. As a transport layer protocol SCTP uses end to end metrics, such as Retransmission Time Out (RTO), to manage mobility handover. Our investigation illustrates that Wireless LAN (WLAN) mobility causes continuously increased Round Trip Times (RTT) resulting from 802.11 MAC retransmissions, regardless of the service specified by upper layers. We present scenarios where the current understanding of SCTP switchover aggressiveness is invalid; spurious failovers together with excessive RTO result in new forms of receiver buffer blocking communication failure. Given wireless mobility performance issues, together with the ambiguity of end to end metrics, we propose an Adaptive Optimized RTO algorithm for wireless Access Networks (AORAN) which uses local as well as end to end metrics to manage mobility. AORAN measures RTT between the mobile node and Access Point (AP) to calculate wireless and Internet RTO subcomponents. We also show binary exponential backoff has negative effects on SCTP with increased wireless RTT; AORAN introduces a decision mechanism which implements backoff on RTO subcomponents only when appropriate.

Keywords: SCTP, mobility, multi-homing, failover, receiver buffer blocking.

1 Introduction

In recent years there has been increasing demand for the provision of a seamless end user experience for real time applications such as Voice Over IP (VOIP). WLAN networks are a critical element of any pervasive heterogeneous network implementation. Limited signal range however, means that any effective WLAN network is achieved through multiple overlapping coverage zones and mobile applications will therefore require multiple switchovers. The transport layer Stream Control Transmission Protocol (SCTP) [1][2][3] provides support for transparent mobility. This is achieved through multi-homing - the ability to implement an end-to-end communication session transparently over multiple physical paths.

In earlier work [4][5] we presented results showing that excessive RTO values for primary paths caused SCTP to behave counter intuitively by significantly delaying path failover. This paper identifies further issues relating to the calculation of RTO for degraded alternate paths, which result in total association communication failure.

H. van den Berg et al. (Eds.): WWIC 2009, LNCS 5546, pp. 133–145, 2009.
© Springer-Verlag Berlin Heidelberg 2009

We illustrate that excessive RTO values result from continually increased Access Network (AN) RTT caused by the division of functionality between 802.2 Logical Link Control (LLC) and the 802.11 Medium Access Control (MAC). The 802.11 MAC retransmits packets regardless of the service type requested by upper layers [6].

Other studies [7] suggested that spurious failovers, those triggered by loss not by path failure, did not degrade system performance as switchback from degraded alternate paths was immediate, even for WLAN. We illustrate that although switchback to the primary path is immediate, excessively large RTO values on the alternate path can result in communication failure of up to 12 seconds.

As a connection oriented protocol SCTP must guarantee in order delivery of packets to the receiving application layer. We illustrate that excessive RTO values delay the detection of packet loss, causing a new form of receiver buffer blocking communication failure. The concept and effects of receiver buffer blocking is described in [8]. Furthermore, we illustrate that the SCTP retransmission strategy, which fast retransmits on the same path and timeout retransmits on an alternate path, also causes a second new form of receiver buffer blocking.

Our solution AORAN, measures mobile node to AP RTT independently of end to end RTT. It introduces local wireless and Internet RTO subcomponents which are combined to calculate end to end RTO. AORAN also implements a decision mechanism which selectively implements Binary Exponential Backoff (BEB) to the RTO subcomponents depending on network conditions. Results presented illustrate that AORAN removes the new forms of receiver buffer blocking and transmits 35%, 45% and 31% more data than standard SCTP for Path Max Retransmissions (PMR) 0 – 2 respectively.

2   Related Work

In previous work [4][5] we indicated that the characteristics of wireless environments caused SCTP performance deficiencies, resulting in delayed path switchover and reduced throughput. In [4] we identified that the standard SCTP RTO calculation mechanism was inappropriate in WLAN environments, since increased RTT significantly distort RTO calculations. It was shown that SCTP behaved in a counterintuitive manner which allowed more time for switchover as network conditions degraded. In [5] we investigated how alterations to the standard SCTP RTO parameters α, the smoothing factor, and β, the delay variance factor could preserve protocol layer boundaries while addressing the performance limitations highlighted in [4]. Results presented indicated that while performance improvements were achievable through careful selection of α and β values, the switchover delays which resulted from the distortions caused by continuously increasing RTT values in WLAN environments persisted. In this paper we propose to address the performance limitations in [4] using a cross layer approach which uses network layer, RTT between the mobile node and AP, as well as transport layer end to end performance characteristics for path performance evaluation.

In [9] an explicit retransmission notification is introduced for use by upper layer congestion control procedures. In [10] it is indicated that local retransmission between the 802.11 MAC and access points distort RTT calculation for TCP. A mechanism is introduced which hides the duration of the 802.11 MAC error recovery phase from upper layer protocols. In [11] it is proposed that link-layer retransmissions result in delays which cause undesired control actions in TCP. Additional delays are inserted in link layer packets which results in the calculation of more appropriate TCP values. These studies are for 802.11 delay spikes for TCP and do not consider SCTP or mobility with degraded Receiver Signal Strength (RSS) for wireless networks.

In [12] the tradeoff between access point buffering and loss for voice traffic in 802.11 networks is investigated. As a result of mobility an abrupt transition from the low-loss, low delay regime to high-loss, high-delay operation is observed. The study indicates that below the point of performance transition, AP buffer size has little impact on throughput. However as the point of performance transition is passed total delay depends strongly on AP buffer size. In this paper we observe this point of performance transition. We suggest however, that 802.11 MAC retransmissions rather than the buffer sizes are the primary cause of increased RTT.

A number of investigations have been undertaken which analyse SCTP switch performance. In [7] it was suggested that aggressive failover strategies achieved through low PMR values did not degrade performance as spurious failovers were temporary in nature and resulted in an immediate switchback. Furthermore, aggressive switching strategies were shown to improve throughput regardless of the alternate paths' characteristics (bandwidth, delay, and loss rate). Results presented here illustrate that when the alternate path has high RTT and loss rates, spurious failovers significantly degrade performance. In [13] it is identified that the 3 tunable parameters RTO.min, RTO.Max and PMR have the dominant effect on SCTP switch detection. In [14] an SCTP handover scheme for VoIP applications is presented. The proposed handover scheme is based on the ITU-T E-Model for voice quality.

For a multi-homed protocol retransmission policy is more complex than for a single-homed protocol. A number of studies have investigated how the characteristics of alternate paths impact on SCTP performance. In [15] different SCTP retransmission policies are investigated for a lossy environment. A retransmission strategy which fast retransmits on the same path and timeout retransmits on an alternate path is suggested. This alteration was accepted as part of the standard. Our results illustrate that the retransmission of timeout packets on alternate degraded paths results in receiver buffer blocking. In [16] the authors identify that a finite receiver buffer will block concurrent SCTP transmission when the quality of one path is lower than others. Several retransmission policies are studied which can alleviate receiver buffer blocking. [17] illustrates that SCTP retransmission performance will degrade acutely when the secondary path delay is more than the primary path delay. In [18] a "potentially-failed" state is introduced for SCTP concurrent path usage to alleviate receiver buffer blocking.

3   An Investigation of Continually Increased Wireless Delay

As a mobile node moves from the coverage of an access point delay increases significantly. This delay is not characteristic of traditional WLAN delay spikes, rather it is recognizable by its significant and continuously increasing nature. We will illustrate in later sections how this negatively impacts on SCTP mobility. In this section we investigate the causes of this increased WLAN RTT.

Through 802.2 LLC the IEEE have defined a uniform interface by which upper layers can access the services of carrier protocols. Below the LLC a MAC sublayer implements packet transmission for specific medium types. There are 3 types of service specified by 802.2; Type 1 an unacknowledged connectionless service, Type 2 connection-oriented and Type 3 is an acknowledged connectionless service.

In order to investigate how MAC layer packet retransmissions affected RTT an experimental test configuration was created which consisted of two Laptops, representing a mobile client and a back end server, connected by two Linksys WRT54GL 802.11g access points. Both laptops were configured with SCTPLIB [19]. The SCTP client was multi-homed using 2 wireless interfaces. The SCTP server was hosted by a Dell D830 laptop. RTO parameters were configured with default values RTO.min = 1s, RTO.max = 60s, RTO.Initial = 3s, PMR was set to 1. Background traffic was generated using 250Kbytes of ICMP data.

Our experimental configuration used LLC Type 1 service, as can be verified by the Type 1 indicator *0x03* in the traced packets. 802.11 MAC implements its own positive frame acknowledgment independently of 802.2 through an 802.11 ACK control frame[6][20]. In this way the 802.11 MAC at the receiving station implements error control independently of 802.2. If the sending station doesn't receive a MAC level ACK after a period of time, it will retransmit the frame.

As the mobile node moves from the coverage area of the AP RSS degrades and results in intermittent network connectivity. Fig1 illustrates the SCTP RTO and SRTT. In [4] (Section 3) it is illustrated how SCTP RTO calculation is dependant on RTT. The continually increased SRTT from approximately 60s results in the calculation of excessive RTO.

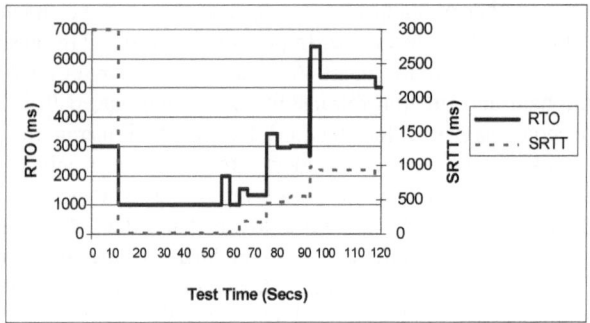

Fig. 1. SCTP SRTT and RTO

In order to determine how intermittent RSS resulted in increased RTT a Wireshark[21] trace using AIRPCAPs 802.11 Wireless Packet capture plug-in[22] was taken. Table 1 details the packet transmissions for a selected period.

At 74.26s SCTP data with TSN 24246 is successfully transmitted. At 74.27s an ICMP packet is transmitted, this packet requires 1 retransmission before it is successfully acknowledged. Since the 802.2 LLC is operating in Type 1 mode this retransmission was handled by the 802.11 MAC. From 74.27s to 74.31s ICMP 7400 is repeatedly, yet unsuccessfully retransmitted by the 802.11 MAC until a retransmission limit is exceeded. Similar retransmissions of 802.11 packets occur until 75.38s when the 802.11 MAC is finally available to SACK SCTP TSN 24246 resulting in an RTT of 1.12s. Fig 2 details the number of 802.11 MAC retransmissions.

Table 1. Wireshark Trace

Time	Source	Destination	Protocol Type	Information
74.26	192.168.2.200	192.168.2.210	SCTP	DATA (TSN = 24246)
74.27	192.168.2.210	192.168.2.2	IP	proto=ICMP 5920
74.27	192.168.2.210	192.168.2.2	IP	proto=ICMP 5920
74.27		HonHaiPr_3d:eb	IEEE 802.11	Acknowledgement
74.27	192.168.2.210	192.168.2.2	IP	proto=ICMP 7400
74.27	192.168.2.210	192.168.2.2	IP	proto=ICMP 7400
74.28	192.168.2.210	192.168.2.2	IP	proto=ICMP 7400
74.29	192.168.2.210	192.168.2.2	IP	proto=ICMP 7400
74.30	192.168.2.210	192.168.2.2	IP	proto=ICMP 7400
74.31	192.168.2.210	192.168.2.2	IP	proto=ICMP 7400
74.31	192.168.2.210	192.168.2.2	IP	proto=ICMP 8880
74.32	192.168.2.210	192.168.2.2	IP	proto=ICMP 8880
74.38	192.168.2.210	192.168.2.2	IP	proto=ICMP 8880
74.42	192.168.2.210	192.168.2.2	IP	proto=ICMP 8880
74.48		HonHaiPr_3d:eb	IEEE 802.11	Acknowledgement
:	:	:	:	:
75.38	192.168.2.210	192.168.2.200	SCTP	SACK (TSN =24246)

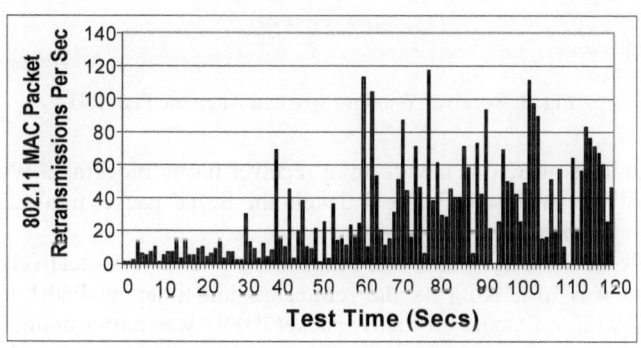

Fig. 2. 802.11 MAC Retransmissions

Fig2 illustrates that as RSS degrades the number of 802.11 MAC retransmissions increases significantly. For the 40s intervals from 0-39s, 40-79s and 80-120s the average number of 802.11 MAC retransmissions were 10.0, 36.8 and 47.8 respectively. The retransmissions at the 802.11 MAC increased the delay between the transmission and acknowledgement of upper layer SCTP packets. We will now illustrate how this results in receiver buffer blocking.

4   Receiver Buffer Blocking Resulting from Excessive Alternate Path RTO

In [4] it is illustrated how SCTP path management is significantly dependant on RTT and subsequent RTO values. Using the test configuration described in Section 3 an SCTP association is established. The SCTP standard specifies that RTO is calculated separately for each path, while a single receiver window is utilized for all paths within the association. Fig3 illustrates the alternate path RTO and receiver window size for the association. At approximately 37s the Receiver Window (rwnd) falls below the size of one packet and again between 39s and 45s resulting in receiver buffer blocking. This new form of receiver buffer blocking prevents packet transmission on all paths. A period of increased RTO from 39s is also identified.

For the 15s intervals 0-14s, 15-29s and 30-44s the number of MAC retransmissions were 21, 50 and 363 illustrating a general degradation in path performance. The effect of this degradation is twofold: it increases baseline RTT (3.4 to 3.7 [4]) as well as resulting in packet loss which doubles RTO (3.8[4]).

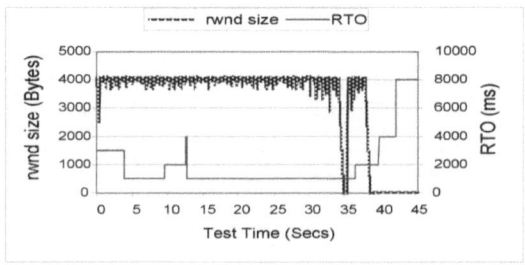

Fig. 3. Receiver Window Size and Alternate Path RTO

In order to investigate this new form of receiver buffer blocking a Wireshark trace for the period was analysed. Figure 4 details the SCTP packet transmissions on the alternate path for the period 39-45s.

At 39.609s the RTO for Path1 and Path2 were 1s and 4s respectively. At 39.609s packet 21893 was lost. After 1s the retransmission timer on Path1 timed out. At 41.250s, following a timeout on Path1, packet 21893 was retransmitted on Path2. At

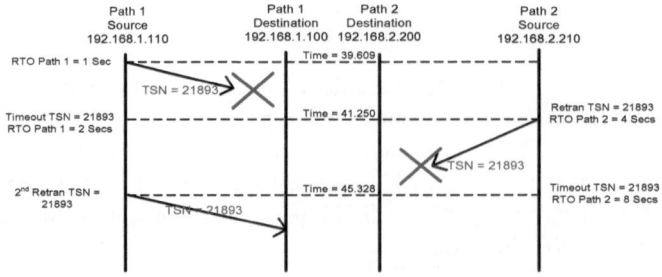

Fig. 4. SCTP Alternate Path Packet Transmission 39-45s

this time Path2 was in a degraded condition and packet 21893 was lost. Since the RTO on Path2 was 4s the loss of packet 21893 was not recognized until 45.250 (41.250+4)s. Packet 21893 is then retransmitted on Path1.

As a connection oriented protocol SCTP must guarantee in sequence delivery of packets to the application layer. Packet 21893 was lost at 39.609s causing the receiver to buffer out of sequence packets until 40.270s. At this time the rwnd fell below the size of 1 packet. Since the loss of the retransmission of packet 21893 on path 2 was not detected until 45.250s the association was stalled for over 5s by the receiver awaiting the retransmission of this lost packet.

Receiver buffer blocking has major implications for SCTP performance as it causes data transmission for the entire association to cease for a period of time. It is primarily an issue when the paths in an SCTP association are concurrently used for data transfer. Our results indicate however, that the characteristics of wireless mobile networking can result in receiver buffer blocking for standard SCTP.

The mechanisms used for addressing receiver buffer blocking generally attempt to reduce its occurrence and lessen its impact [16][17][18]. We propose an adaptive RTO calculation and retransmission mechanism for SCTP. The SCTP RTO mechanism inherited from TCP assumes that packet loss is as a result of Internet congestion. Therefore, to avoid network overload Binary Exponential Back-off (BEB) is initiated. The receiver buffer blocking descried is as a result of packet loss resulting from wireless AN performance deficiencies rather than Internet congestion. In a degraded AN where packet loss occurs with high baseline RTT resulting from 802.11 retransmissions, the implementation of RTO doubling results in excessively large values which are not appropriate to actual network conditions. The following sections analyse how the aggressiveness of mobility coupled with an adaptive RTO calculation and retransmission mechanism improve SCTP performance.

5   Receiver Buffer Blocking Resulting from Spurious Failover

In order to evaluate the effect of aggressive path selection an NS2 [23] simulation which utilizes the University of Delawares [24] SCTP module was created. Simulations were ran with PMR values ranging from 5, least aggressive, to 0, most aggressive. Fig5 illustrates the accumulated data transmitted for each PMR value.

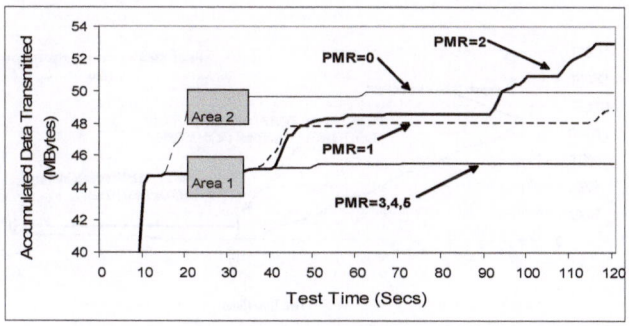

Fig. 5. Accumulated Data Transmitted by PMR

The receiver buffer blocking described in Section 4 is highlighted by Area 1 in Fig5 from 21-34s. In Area 2 in Fig5 there is an association communication failure between 22s and 35s for PMR 0. For PMR=0 packet loss results in path failover, not packet retransmission, therefore this communication failure is not as a result of the form of receiver buffer blocking described in Section 4. Fig6 illustrates the sequence of packet transmissions for Area 2.

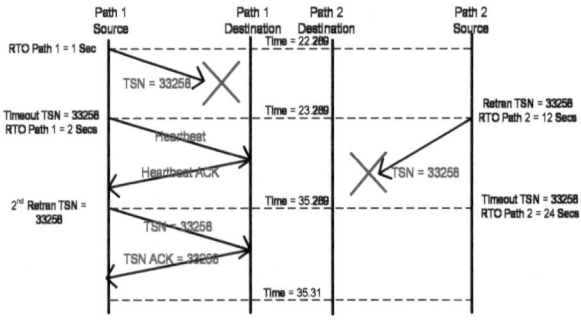

Fig. 6. Packet Transmission for 22-35s (PMR=0)

At 22.289s the RTO for Path1 and Path2 were 1s and 12s respectively. At 22.289s packet 33258 was lost. After 1s the retransmission timer on Path1 timed out. As PMR=0 path failover occurs from Path1 to Path2. Path1 is marked INACTIVE and a heartbeat is transmitted. A heartbeat acknowledgment is returned and Path1 is set to ACTIVE. SCTP then sets an indication that subsequent packets should be sent on the reactivated primary path.

At 23.289s, following a timeout on Path1, packet 33258 was retransmitted on Path2. At this time Path2 according to Fig1 was in a degraded condition and packet 33258 was lost. Since the RTO on Path2 was 12s the loss of packet 33258 was not recognized until 35.289(23.289+12)s. Packet 33258 is then successfully retransmitted on Path1. During the period 22.289 – 35.289s no data was transmitted by the association. In order to further explain this communication failure Fig7 illustrates the rwnd for the period 22s to 35s.

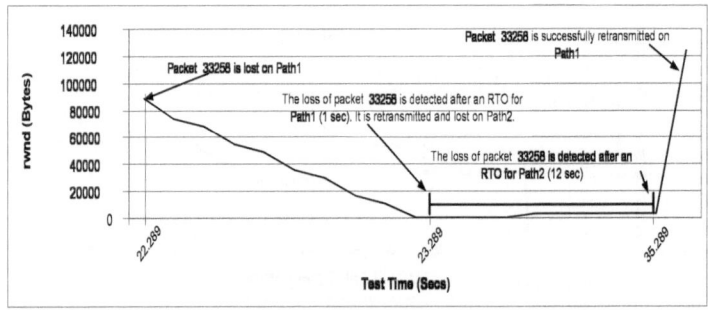

Fig. 7. Receiver Window for 22-35s (PMR=0)

Packet 33258 was lost at 22.289s. The receiver buffered out of sequence packets until 23.289s when the rwnd fell below the size of 1 packet. Since the loss of packet 33258 was not detected until 35.289s the association was stalled by the receiver awaiting this lost packet.

Fig7 therefore illustrates another new form of receiver buffer blocking for PMR=0. In [7] it was suggested that aggressive failover strategies achieved through low PMR values did not degrade performance as spurious failovers were temporary in nature. Furthermore, aggressive switching strategies were shown to improve throughput regardless of the alternate paths' characteristics (bandwidth, delay, and loss rate). Our results illustrate that in a wireless environment spurious failover can result in receiver buffer blocking.

While the spurious failover resulted in receiver buffer blocking, the excessive RTO value of 12s for the alternate path compounded the performance deficiency. The 12s RTO was calculated as a result of repeated packet loss and implementation of BEB. In the following section we investigate a more select implementation of BEB.

6   Proposed Alterations to the SCTP Protocol

For a connection oriented protocol such as TCP, RTO calculation is a tradeoff between retransmission delays and spurious retransmission of valid packets. In addition to these tradeoffs SCTP must also consider the effect of RTO on mobility. Previous sections illustrated how excessive RTO resulted in 2 new forms of receiver buffer blocking. We propose the AORAN algorithm to address these performance deficiencies.

6.1   The AORAN Algorithm

Outlined below is an adaptive RTO calculation mechanism. A parameter ANRTT, which records local delay between the mobile node and access point is introduced. An estimation of Internet RTT is calculated by subtracting ANRTT from end2endRTT. RTO consists of 2 subcomponents, one reflecting the condition of the AN and one reflecting the condition of the Internet. A number of studies have been undertaken which propose mechanisms by which Internet congestion can be recognized [25][26]. For SCTP a draft standard[27] proposes the addition of an Explicit Congestion Notification (ECN) packet. We introduce inetcongst an indicaton of Internet congestion which is controlled by ECN and eANRTT an indication of increased AN RTT. For each subcomponent the calculation of RTO is implemented using (3.1) to (3.7) from [4]. The calculation of RTO as a result of packet loss and the implementation of BEB (3.8) is now selectively implemented by AORAN.

```
bool inetcongst //set if ECN chunk is received
bool eANRTT //set if exponential change in AN RTT

Event::Packet Arrival
        end2endRTT = time sent - time received
        inetRTT = end2endRTT-ANRTT //Internet component of RTT
        inetRTO = inetSRTT+4 * inetRTTVAR
        inetRTTVAR.new =   (1-β) * inetRTTVAR.old+.β * (inetSRTT.old- inetRTT.new)
        inetSRTT.new = (1-.α) * inetSRTT.old+.α * inetRTT.new
```

```
        ANRTO = ANSRTT+4 * ANRTTVAR
        ANRTTVAR.new =    (1-β) * ANRTTVAR.old+.β * (ANSRTT.old- ANRTT.new)
        ANSRTT.new = (1-.α) * ANSRTT.old+.α * ANRTT.new
   RTO = ANRTO + InternetRTO // Total RTO for path
Event::Packet Loss
   PMRCounter++ // Increment retransmission counter
   if(PMRCounter> PMRLimit)
        ActivePathID = AlternatePathID[0] // Switch to alternate path
   else
        Retransmit Packet (ActivePathID) // Retransmit on same path
        if(eANRTT==TRUE && inetcongst==TRUE)
             inetRTO = inetRTO *2 //BEB
             ANRTO = ANRTO
        ELSE if(eANRTT==TRUE && inetcongst==FALSE)
             inetRTO = inetRTO
             ANRTO = ANRTO
        ELSE if(eANRTT==FALSE && inetcongst==TRUE)
             inetRTO = inetRTO *2 //BEB
             ANRTO = ANRTO*2 //BEB
        RTO = ANRTO + InternetRTO
```

6.2 An Evaluation of the AORAN Algorithm

An NS2 simulation model including WLAN access and Internet backbone is illustrated in Fig8. Node S and R are SCTP endpoints. Internet congestion is introduced
by the TCP sending nodes which transmit at a rate of 1Gbps across the backbone to
the TCP receiving nodes.

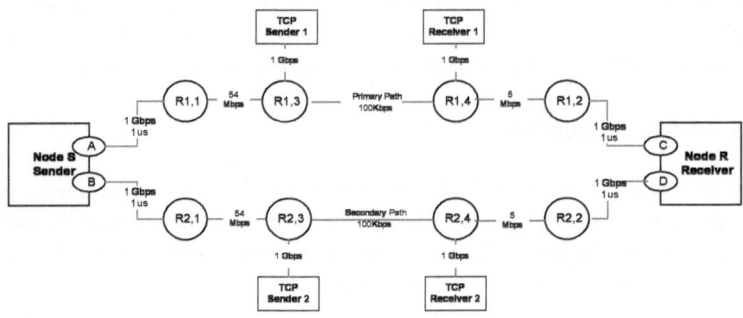

Fig. 8. NS2 Simulation including Internet Backbone

We initially evaluate the performance of AORAN with increased AN RTT and
without Internet congestion. The performance of the strategy in the presence of Internet congestion is discussed later. Fig9 compares the performance of AORAN against
standard SCTP for PMR values ranging from 0-2. The simulations with PMR values
3-5 have similar performance to PMR=2.

Section 4 identified a new form of receiver buffer blocking resulting from the retransmission of packets on degraded alternate paths. Section 5 illustrated that aggressive path failover with PMR=0 also caused a second new form of receiver buffer
blocking following spurious failover. Fig 9 illustrates that AORAN has removed these
forms of receiver buffer blocking.

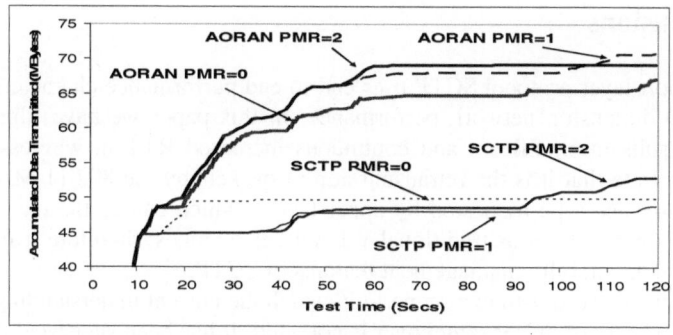

Fig. 9. Accumulated Data Transmitted AORAN Vs SCTP without Internet Congestion

Fig9 and Table 2 indicate that AORAN transmitted 35%, 45% and 31% more data than standard SCTP for PMR 0 – 2 respectively. As stated earlier the current understanding for SCTP mobility suggests that low PMR values improve throughput even for wireless environments. The results presented in Fig9 and Table 2 indicate that more moderate switching strategies with larger PMR values (e.g. 1 or 2) have better performance than PMR=0.

Table 2. Accumulated Data Transmitted for the Association

	PMR=0	PMR=1	PMR=2
AORAN	67.3	71.0	69.2
Standard SCTP	50.1	48.8	52.9

Fig10 illustrates that AORAN is compliant with Internet congestion policy. When Internet congestion occurred without increased AN RTT AORAN and SCTP behaved identically. When there is both Internet congestion and increased AN RTT AORAN transmits similar amounts of data to SCTP.

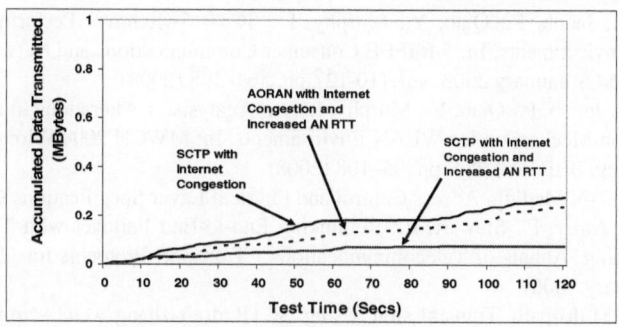

Fig. 10. Accumulated Data Transmitted AORAN Vs SCTP with Internet Congestion

7   Conclusions

As a transport layer protocol SCTP uses end to end performance characteristics, such as RTO, to dimension network performance. In this paper we have illustrated that mobility results in significant and continuous increased RTT in wireless networks. Results illustrate that it is the retransmission of packets by the 802.11 MAC, regardless of the service type requested by upper layers, which causes the increased RTT. The delay is not as brief as traditional WLAN delay spikes; therefore it impacts significantly on the mobility management features of SCTP.

Results presented illustrate a scenario in which the current understanding regarding the aggressiveness of SCTP switchover is not valid. It has been suggested that aggressive failover improves throughput regardless of the alternate paths characteristics. We illustrate that spurious failovers together with excessive RTO can result in receiver buffer blocking communication failure.

Given the performance issues introduced by wireless mobility, coupled with the ambiguity of end to end metrics, we propose an Adaptive Optimized RTO algorithm for wireless Access Networks (AORAN). AORAN explicitly measures RTT between the mobile node and AP. This RTT is used to calculate a local wireless RTO which is used as a component in end to end RTO. We have also shown that the implementation of binary exponential backoff in the presence of increased AN RTT has an exponentially negative effect on SCTP performance. Therefore AORAN selectively implements binary exponential backoff only as appropriate. Results presented illustrate that AORAN removes the new forms of receiver buffer blocking and transmits significantly more data than standard SCTP.

References

[1] Stewart, R., Xie, Q., et al.: Stream Control Transmission Protocol, RFC 4960 (September 2007)
[2] Stewart, R., Xie, Q., et al.: Stream Control Transmission Protocol, Dynamic Address Reconfiguration, proposed extension (May 2006)
[3] Riegel, M., Tuexen, M.: Mobile SCTP, proposed draft (March 2006)
[4] Fallon, S., Jacob, P., Qiao, Y., Murphy, L.: SCTP Switchover Performance Issues in WLAN Environments. In: 5th IEEE Consumer Communications and Networking Conference (CCNC), January 2008, vol. (10-12), pp. 564–568 (2008)
[5] Fallon, S., Jacob, P., Qiao, Y., Murphy, L.: An Analysis of Alterations to the SCTP RTO Calculation Mechanism for WLAN Environments. In: MWCN 2008 Wireless and Mobile Networking. IFIP, vol. 284, pp. 95–108 (2008)
[6] Wireless LAN Medium Access Control and Physical Layer Specifications (March 2007)
[7] Caro, A., Amer, P., Stewart, R.: Rethinking End-to-End Failover with Transport Layer Multihoming. Annals of Telecommunications - Transport Protocols for NGNs 61 (January-February 2006)
[8] Wireless Multi-path Transmission Using SCTP draft-zhang-wcmt-sctp-00 (September 2008)
[9] Dedu, E., Linck, S., Spies, F.: Removing the MAC Retransmissions Times from the RTT in TCP. In: Euromedia Conference 2005 (2005)

[10] Ratnam, K., Matta, I.: Effect of local retransmission at wireless access points on the round trip time estimation of TCP. In: 31st Annual Simulation Symposium 1998 (1998)

[11] Moller, N., Johansson, K.H., Hjalmarsson, H.: Making retransmission delays in wireless links friendlier to TCP. In: 43rd IEEE Conference on Decision and Control (CDC), vol. 5(14-17), pp. 5134–5139 (December 2004)

[12] Malone, D.W., Clifford, P., Leith, D.J.: On Buffer Sizing for Voice in 802.11 WLANs. IEEE Communications Letters 10(10), 701–703 (2006)

[13] Caro Jr., A.L., Amer, P.D., Stewart, R.R.: End-to-end failover thresholds for transport layer multihoming. In: MILCOM 2004, vol. 1, pp. 99–105. IEEE, Los Alamitos (2004)

[14] Fitzpatrick, J., Murphy, S., Murphy, J.: An Approach to Transport Layer Handover of VoIP over WLAN. In: 3rd IEEE Consumer Communications and Networking Conference (CCNC), vol. 2, pp. 1093–1097 (2006)

[15] Caro, A., Amer, P., Stewart, R.: Retransmission Schemes for End- to-end Failover with Transport Layer Multihoming. In: GLOBECOMM 2004 (2004)

[16] Iyengar, J., Amer, P., Stewart, R.: Receive buffer blocking in concurrent multipath transfer. In: Globecom 2005, vol. 3, pp. 1341–1347 (2005)

[17] Qiao, Y., Fallon, E., Murphy, J., Murphy, L.: SCTP performance issue on path delay differential. In: Boavida, F., Monteiro, E., Mascolo, S., Koucheryavy, Y. (eds.) WWIC 2007. LNCS, vol. 4517, pp. 43–54. Springer, Heidelberg (2007)

[18] Natarajan, P., Shah, K., Amer, P.: Concurrent Multipath Transfer using SCTP Multihoming: Introducing the Potentially-failed Destination State. In: IFIP Networking 2008 (2008)

[19] SCTP library (sctplib), version sctplib-1.0.5, http://www.sctp.de

[20] 802.11 WLAN Hands-on Analysis - Unleashing the Network Monitor for troubleshooting and Optimization. Byron W Putman

[21] Combs, G., et al.: Wireshark network protocol Analyzer, Version 0.99.5

[22] AirPcap 802.11 Wireless Packet Capture CACE Technologies

[23] UC Berkeley, LBL, USC/ISI, and Xerox Parc: ns-2 Version 2.29 (October 2005)

[24] Caro, A., et al.: ns-2 SCTP module, Version 3.5

[25] Floyd, S.: TCP and Explicit Congestion Notification. ACM Computer Communication Review 24 (1994)

[26] Kuzmanovic, A.: The Power of Explicit Congestion Notification. ACM SIGCOMM Computer Communication Review 35(4) (2005)

[27] Ladha, S., et al.: draft-ladha-sctp-nonce-06 (January 2007)

Handover Incentives for WLANs
with Overlapping Coverage*

Xenofon Fafoutis and Vasilios A. Siris

Institute of Computer Science (ICS)
Foundation for Research and Technology - Hellas (FORTH)
P.O. Box 1385, GR 711 10 Heraklion, Crete, Greece
{fontas,vsiris}@ics.forth.gr

Abstract. It is well known that in IEEE 802.11 networks, the assignment of low-rate and high-rate users to the same access point significantly degrades the performance of the high-rate users. Our objective is to investigate the implications of the above performance degradation on the incentives for handover between 802.11 wireless local area networks with overlapping coverage. Our focus is on the incentives for supporting handovers, due solely to the improved performance handovers yield for both wireless networks. To study the phenomenon and estimate the potential gain of such handovers, we propose a simple model that predicts the throughput of each access point in different cases. The throughput approximation model can indicate when the handover is expected to be beneficial, and can be used in a handover acceptance policy. Simulation of the proposed procedure suggests that the model is accurate and that there are significant throughput gains for both wireless networks.

Keywords: handovers, cooperation incentives, wireless access network.

1 Introduction

It is well known that in IEEE 802.11 networks, the assignment of low-rate and high-rate users to the same access point significantly degrades the performance of the high-rate users [1]. This occurs because IEEE 802.11's medium access control protocol gives both high and low-rate nodes equal chances for accessing the shared wireless channel. However, low-rate nodes need more time to send the same amount of data. As a result, high-rate users suffer significant performance degradation, achieving throughput equal to that of low-rate users. When using the term *rate*, we refer to the modulation rate of an 802.11 transmitter.

The objective of the work presented in this paper is to investigate the implications of the above performance degradation on the incentives for handover between 802.11 wireless local area networks with overlapping coverage. In case

* This work was supported in part by the European Commission in the 7th Framework Programme through project EU-MESH (Enhanced, Ubiquitous, and Dependable Broadband Access using MESH Networks), ICT-215320, http://www.eu-mesh.eu

H. van den Berg et al. (Eds.): WWIC 2009, LNCS 5546, pp. 146–158, 2009.

of operator-owned networks, handovers may be supported by cooperation agreements between the operators. The focus of this work is on the incentives for supporting handovers, due solely to the improved performance that handovers yield for both wireless networks, without involving any monetary exchange. Moreover, similar performance incentives arise in the case of overlapping wireless home networks where cooperation agreements are unrealistic. The main assumption of these scenarios is that two or more access points operate in the same channel. Indeed, it is common that there are more than three access points within the range of each other [2]. Hence, the three orthogonal channels available in 802.11b and 802.11g are not sufficient to assign orthogonal channels to different access points. Moreover, as more wireless networks operating in unlicensed bands are deployed over time, the above scenario will be more dominant.

In order to study the phenomenon and measure the potential gain of the aforementioned handovers, we propose a simple model which predicts the throughput of each access point in different context. Similar throughput approximations have been used in [9], [10], [11], [12] and [13]. Our main contribution lies in the application of the above model to predict the impact of handovers and decide whether handing over the nodes that communicate at low rates is beneficial for both access points. Our model suggests that there are cases where it is highly probable that the performance of both wireless networks is significantly improved. For instance, in 802.11b and when the low-rate nodes transmit at 1Mbps, our model suggests that if the number of nodes is uniformly distributed over time there is a 74% probability that the handover is beneficial for both access points, providing, on average, 55% better throughput for the nodes of the low-rate nodes' new host and more than three times better throughput for the nodes of their initial access point. Additionally, based on the necessary conditions for a handover to be beneficial for both access points, we propose a policy that can be used by each access point in order to decide whether to cooperate to the handover.

The rest of this paper is organized as follows. Section 2 summarizes the related work. Section 3 describes the scenario we investigate, and presents the throughput model used in the analysis. Section 4 quantifies the potential gains according to the model. Section 5 compares the analytical model with simulation using NS-2. Section 6 extends the model to include alternative bottlenecks, such as ADSL connections, and investigates their influence. Finally, Section 7 concludes the paper and presents our future work on this subject.

2   Related Work

The handover incentives studied in this paper address the performance degradation problem of 802.11 networks from a new perspective. Related work has approached the same issue in different ways.

One approach to mitigate the problem is to reduce the time low-rate transmissions utilize the shared medium, thus providing time fairness [3][4]. Although such an approach can increase the aggregate throughput of the network, it is unfair to low-rate nodes, as it minimizes their throughput even more. This leads

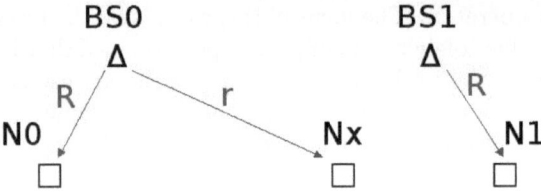

Fig. 1. BS0 serves both high-rate (N0) and low-rate (Nx) users (case a)

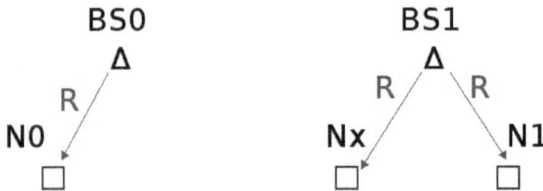

Fig. 2. BS0's low-rate users (Nx) are handed-off to BS1 (case b)

to unsatisfied clients, which should be avoided in an environment with multiple competing providers. Another approach makes use of relay nodes for transmissions that cannot be performed at high rates [5][6][8]. In [5], the authors replace low rate transmissions with a two-hop sequence of shorter range, that enable higher rate transmissions. In [6], the authors propose a system which opportunistically transforms higher-rate 802.11 stations into relays for stations with low data-rates, hence requires the availability of such relay nodes. Another approach is to aggregate the capacity of all the access points and use load balancing mechanisms in order to maximize the network performance [7]. This approach requires all access points to cooperate, which can be assumed in the case where the access points belong to the same operator, but not when they belong to different operators. On the other hand, the focus of this paper is exactly on the cooperation between different operators, and shows that such cooperation can result from performance-oriented incentives even when operators act in their own self-interest.

3   Throughput Model

Consider the case of two access points, BS0 and BS1, Fig. 1. BS0 sends traffic to N0 nodes at high rate R and to Nx nodes at low rate r. Nodes N0 and Nx are the clients of BS0 and its actions target to improve its clients throughput. BS1 sends traffic to nodes N1 at high rate R. Nodes Nx are closer to BS1 and would transmit at a higher rate R, if they were associated to it. This is the base scenario, which we will refer to as case a. Additionally, the following assumptions are made: (a) both access points operate at the same channel, (b) all access points and nodes are in the same contention area and (c) there is at least one node in each of the three node sets. Due to the low-rate transmissions to the set of Nx nodes, the performance for all N0 and N1 nodes degrades.

In the scenario shown in Fig. 2, which we will refer to as case b, the low-rate clients Nx of BS0 are handed-off to BS1. Now, BS1 sends traffic only to the N0 nodes at high rate R, while BS1 sends traffic at high rate R to both its own clients (N1) and the ex-low-rate clients of BS0 (Nx).

The throughput gain of BSi is defined as the ratio of the aggregate throughput of the clients of BSi in case b (Nx clients associated to BS1), over the throughput in case a (Nx clients associated to BS0). This metric will be used to evaluate when the handover of low-rate users (case b) is beneficial. When the gain for both access points is greater than 1, handover improves the aggregate throughput for the clients of both access points. It is important to note that the Nx nodes are clients of BS0 in both cases, even though in the second case they are associated to BS1. Hence, to estimate the gain in both cases, the throughput of the Nx nodes is added to the aggregate throughput of BS0 clients.

The analysis in the following sections focuses on the above simple model, which encompasses the key tradeoffs we want to highlight. The analytical model can be extended to more complex cases, such as more than two access points and the case where additionally to the Nx nodes, some of BS1's clients symmetrically operate at low rates.

3.1   Model

Next we present a model for the throughput in saturated conditions for the downlink direction, i.e., from the access points to the clients. We assume that each access point sends one packet in each round. This is not absolutely true for IEEE 802.11 since the backoff waiting time of each transmission, as defined by the collision avoidance mechanism, is decided probabilistically. However, assuming that the DCF protocol of IEEE 802.11 provides long term fair channel access, the access points will send an equal amount of packets over a long time interval.

If T_0 and T_1 is the time each access point needs to transmit one packet respectively, then the long term throughput in bits per second that each access point will achieve, assuming both access points transmit packets of the same size, is equal to

$$X = \frac{pkt}{T_0 + T_1 + oh} \tag{1}$$

where pkt is the packet size in bits and oh is the overhead of two transmissions.

According to the DCF protocol of IEEE 802.11, each transmitter needs to sense the medium idle for a time interval equal to DIFS. It then chooses a random number of time slots between zero and the contention window (CW) and waits an additional time interval in order to avoid collisions. The transmission follows on the condition that the medium is idle during the time that it is waiting. After the transmission is completed the receiver waits for a time interval equal to SIFS, during which it switches to transmitter mode, and starts transmitting an acknowledgement at the control rate. The duration of DIFS, SIFS and TimeSlot as well as the initial minimum contention window are defined by the protocol. The time interval of the acknowledgement transmission depends on its size and the transmission rate. Since each transmitter chooses a uniformly random number in the interval between zero and CW, the long-term expected overhead of the collision avoidance mechanism when we have only one transmitter, is equal to CW/2 time slots. On the other hand, if a transmitter senses another transmission while being in backoff, it stops his counter and defers until the transmission is over and the medium is idle again, at which point it continues the backoff procedure at the point the backoff was stopped. When there is more than one contending nodes, their backoff counter runs down simultaneously. Hence, the expected overhead due to the contention avoidance mechanism is again CW/2 time slots, assuming that there are no collisions. In the scenario we investigate, there are two contending transmitters. As a result, there is a collision contention probability equal to 0.0625, assuming the minimum contention window is 15, which is the default value for IEEE 802.11. Based on the above, the overhead of two contenting transmissions is

$$oh = 2(DIFS + SIFS + ACK) + \frac{CW}{2}TimeSlot \,, \tag{2}$$

where the values of $DIFS, SIFS, ACK$, and $TimeSlot$ are defined by the 802.11 standard. This overhead ignores potential collisions; we investigate the accuracy of the model in Section 5.

Case a (no handover): When Nx nodes are assigned to BS0, the expected time interval T_0^a that BS0 needs to transmit a packet depends on the percentage of traffic sent to N0 and Nx nodes, since the duration of the transmission is different due to the different rates. On the other hand, the expected time interval T_1^a that BS1 needs to transmit a packet is independent of the number of its nodes since all operate at the same rate. Hence,

$$T_0^a = \frac{N0}{N0 + Nx}\frac{pkt}{R} + \frac{Nx}{N0 + Nx}\frac{pkt}{r} \,, \qquad T_1^a = \frac{pkt}{R} \,, \tag{3}$$

where $N0$ and Nx are the number of nodes in the N0 and Nx node-set respectively, r and R are the low and high rate, respectively.

The expected throughput of each N0 or Nx node, and each N1 node is

$$X_0^a = X_x^a = \frac{1}{N0 + Nx}X^a \,, \qquad X_1^a = \frac{1}{N1}X^a \,, \tag{4}$$

where X^a is estimated from (1), (2), and (3).

Case b (handover of low rate users): In the case there are no low rate transmissions, hence for both access points the expected duration of a packet transmission is equal to

$$T_0^b = T_1^b = \frac{pkt}{R} \,. \tag{5}$$

The expected throughput of each N0 node, and each Nx or N1 node is

$$X_0^b = \frac{1}{N0}X^b \,, \qquad X_x^b = X_1^b = \frac{1}{Nx + N1}X^b \,. \tag{6}$$

The gain of each access point is calculated using the above expressions. When low-rate nodes are associated to BS0 (case a), throughput is reduced. On the other hand, when the low-rate nodes are associated with BS1 (case b), BS1 shares its share of the wireless channel with the Nx nodes, which are BS0's clients. Case b is always beneficial for BS0, since BS0's clients utilize the wireless channel for more than half of the time in case b, and there are improvements due to removing low-rate transmissions. The inequalities

$$GainBS_0 = \frac{X^b + \frac{Nx}{Nx+N1}X^b}{X^a} > 1 \ \text{ and } \ GainBS_1 = \frac{\frac{N1}{Nx+N1}X^b}{X^a} > 1 \quad (7)$$

are necessary conditions for case b to be beneficial for both access points. Since $X^b > X^a$, the $GainBS_0 > 1$ is always satisfied. However, if the assumption that $N0 > 0$ does not hold, then case b is not always beneficial for BS0. The second inequality $GainBS_1 > 1$ is equivalent to

$$\frac{N1}{Nx+N0} > c, \ \text{ where } \ c = \frac{2 + \frac{ohR}{pkt}}{\frac{R}{r} - 1} . \quad (8)$$

This inequality can be used by BS1 to decide if it is beneficial to serve the low-rate nodes of his neighboring access point BS0. We make the following two interesting remarks regarding the above constraint. First, the acceptance constraint does not depend on the ratio of high-rate to low-rate nodes of BS0, but depends only on their sum. Second, for the same R/r ratio a higher value of R yields a higher constraint c. As a result, we expect the constraint to be more restricting for 802.11g than for 802.11b.

By slightly changing the model assuming that the set of Nx nodes consists of two distinct subsets, Ny and Nz, it can be easily shown that given that the handover is beneficial, the gain where both Ny and Nz subsets are served by BS1 is always greater than the gain where the Ny nodes stay with their access point and only the Nz nodes are served by BS1. In fact, the impact of one low-rate node on the throughput is much greater than the impact of additional low-rate nodes. As a result, either all low-rate nodes should be accepted by BS1 or none.

The handover acceptance policy, expressed by (8), requires that the access points know the number of connected nodes and their rates. Assuming that there are no hidden nodes, this information is easy to obtain by sniffing the neighboring traffic. The access point can count unique MAC addresses and extract the rates from their PLCP header. We plan to investigate the implementation details in future work.

4   Model Results

Based on the model presented in the previous section, next we present results to evaluate the number of cases where a handover is advantageous for both access points, and quantify the corresponding throughput improvements. We note that we use packet size equal to 1500 bytes for all the following experiments.

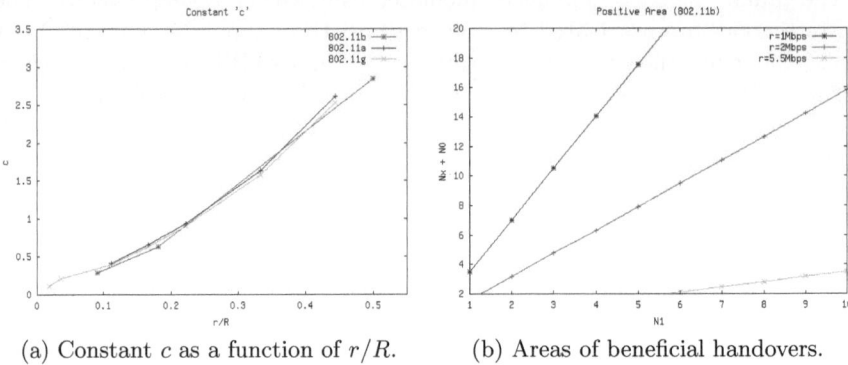

(a) Constant c as a function of r/R.   (b) Areas of beneficial handovers.

Fig. 3. Visualization of (8)

Figure 3(a) depicts the value of constant c, as a function of the ratio r/R, for different versions of 802.11, namely 802.11b, 802.11a, and 802.11g; the high rate R is taken to be the highest rate supported by each version. Since for a positive gain scenario the ratio of BS1 to BS0 nodes needs to be greater than this constant, it is obvious that a higher value of c is more restricting. Moreover, since the high rate R is constant for each version, higher values of r make the handover less beneficial. Obviously, the negative impact of low-rate nodes on the throughput is smaller for higher values of r. As expected, the constant is more restricting for 802.11a/g that operate at 54 Mbps. Additionally, 802.11g is slightly less restricting than 802.11a, because the overhead of the latter is slightly higher.

Figure 3(b) depicts the line defined by (8) for 802.11b and its supported rates. The x and y axis shows the number of BS1 and BS0 nodes, respectively. The slope of each line is given by the constant c. Every $N0, Nx, N1$ combination that is below each line can benefit with the handover of low-rate nodes. The area below each line provides a visual estimation of the probability of the appearance of a beneficial scenario, across all possible scenarios that correspond to all $N0, Nx, N1$ combinations. The figure shows that when the low rate is 1 Mbps beneficial scenarios are highly likely. When the low rate is 2 Mbps the beneficial area is cut in half. When the low rate is 5.5 Mbps the beneficial area is very small, and handovers in this case does not improve the throughput for both access points.

Figure 4 depicts the percentage of scenarios where both access points benefit from the handover. We assume that each scenario is a combination of $N0, Nx, N1$, where each set has a uniformly distributed number of nodes in the interval $[1, 10]$. The total amount of different node combinations is equal to 1000. For 802.11b, when the low rate is equal to 1 Mbps the percentage of beneficial scenarios is almost 74%. At 2 Mbps the percentage drops to 40% and at 5.5Mbps the percentage is about 1%. IEEE 802.11a and 802.11g have similar behavior although the percentages are lower, as expected. The figure indicates that when the low rate is less or equal to 12 Mbps there is a significant percentage (more than approximately

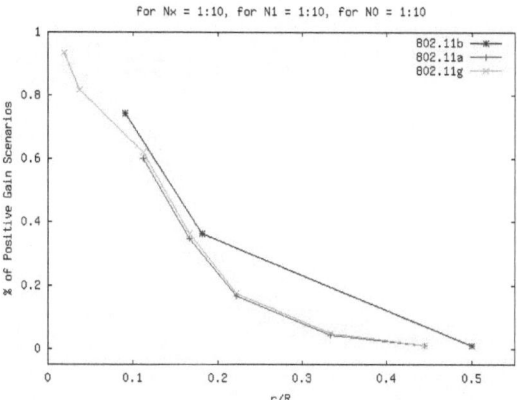

Fig. 4. Percentage of beneficial scenarios

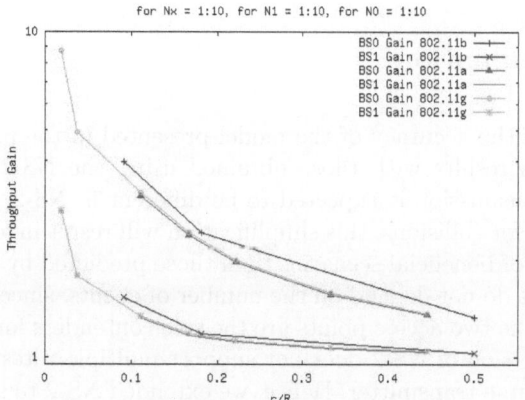

Fig. 5. Throughput gains for BS0 and BS1

35%) of beneficial scenarios. The assumption that number of nodes in each set is uniformly distributed in the interval $[1, 10]$ is, indeed, a simplification. In the real world, the distribution of the number of nodes is expected to be more complicated. However, this simplification provides an estimation on the probability of appearance of a beneficial scenario. Moreover, it can be easily shown that similar conclusions hold for a larger number of nodes.

Figure 5 shows that for higher values of the low rate r, the average gain of the beneficial scenarios decreases. The average gain of BS1, for low rate equal to 1 Mbps in 802.11b is about 55%. At 2 Mbps the gain drops to 20%. Observe that the average gain of BS0 is significantly higher than the gain of BS1; this is because BS0's low-rate clients use a part of BS1's channel access time, when they are associated to the latter.

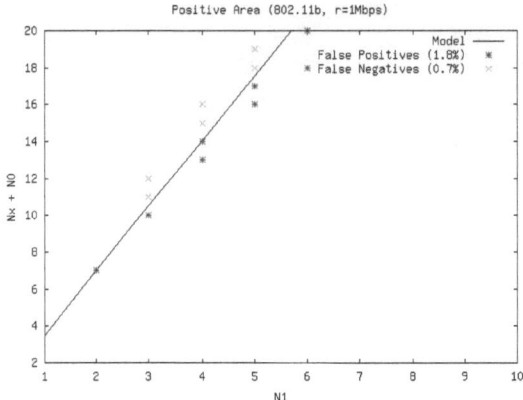

Fig. 6. Comparison of the line with slope c predicted by the model for identifying positive gain scenarios with simulation

5   Model Evaluation

Next we evaluate the accuracy of the model presented in the previous sections, by comparing its results with those obtained using the NS-2 simulator. The overhead of a transmission is expected to be different in NS-2, since the model does not account for collisions; this simplification will result in lower gains and a lower percentage of beneficial scenarios than those predicted by the model. Note that the collisions do not depend on the number of clients, since we focus on the downlink where the two access points are the sole contenders for channel access.

The default version of NS-2 does not support multiple rates for different receivers and the same transmitter. Hence, we extended NS-2 to support multiple rates, by adapting the rate according to the signal strength of the last received packets. The experiments considered 802.11b, and low rate equal to 1 Mbps.

Figure 6 indicates that with simulation, there is no single threshold that separates the positive scenarios from the negative ones. If the constant c predicted by the model is used for accepting or rejecting handovers, there are 1.8% false positives and 0.7% false negatives, which are shown in the figure as red stars and green crosses, respectively. The false positives and false negatives are very close to the threshold line, and their density is higher for a higher number of nodes.

Figure 7 shows the percentage of false positives and false negatives as a function of the acceptance threshold c. The green line vertical line corresponds to the threshold c predicted by the analytical model, which is equal to 0.285. A threshold equal to 0.287 provides a balance of false positives and false negatives. A more conservative threshold equal to 0.301 yields very few false positives, while a threshold equal to 0.34 yields no false positives.

Next we consider the normalized gain, which is a function of the acceptance probability (number of accepted scenarios to the total number of scenarios), AP, and the average gain of the accepted scenarios, $AvgGain$:

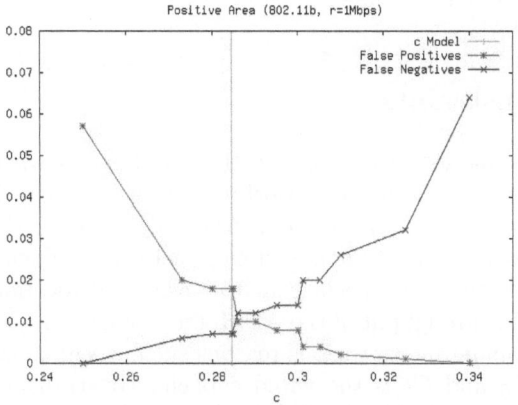

Fig. 7. Percentage of false positives and false negatives for different values of c

$$NormGain = AP \cdot AvgGain + (1 - AP) \cdot 1 \qquad (9)$$

The normalized gain is a metric that takes into account both the probability of appearance and the gain of a beneficial handover and can be used to evaluate the long term gain of an access point that accepts handovers according to our policy. Figure 8 depicts the impact of false positives and false negatives, for different values of c, on the normalized gain. The straight blue horizontal line is a theoretical optimum filter that perfectly predicts the positive and negative scenarios. This figure shows that the acceptance policy based on the threshold c performs extremely well, giving a normalized gain which is within 0.2% of the maximum gain, which is achieved by the theoretical optimal filter. Moreover, Figure 8 shows that the normalized gain predicted by the model (green '+' at gain of approximately 1.39) differs from the normalized gain estimated using

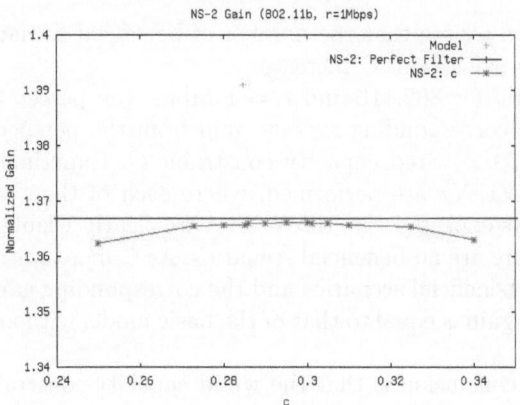

Fig. 8. Normalized gain for different values of c

NS-2 by less than 2.5%, indicating that the model is very accurate in predicting the handover gain.

6   ADSL Constraints

Next we extend the model presented in Section 3 to the case where the bottleneck from a wireless user to the fixed network is not the wireless link, but a wired link after the access point, e.g., a user's ADSL connection. For instance, with modern infrastructure in many home wireless networks, which most commonly use IEEE 802.11g, the bottleneck is in the ADSL connection. Although the theoretic maximum throughput of the ADSL technology is 24 Mbps, much lower throughput is common due to various reasons such as signal attenuation.

Assume that C_0 and C_1 is the wired capacity constraint of BS0 and BS1, respectively. In the remainder of this section, due to space limitations, we consider the simple case where both access points have the same wired capacity constraint, $C_1 = C_2 = C$.

Assume XC^a and XC^b is the aggregate throughput of each access point when there is no handover (case a), and when there is handover (case b), respectively, as predicted by the basic model in Section 3. Since case a includes at least one low rate transmission, whereas case b includes none, it is obvious that $XC^a < XC^b$. As a result, there are three zones where the impact of the wired capacity constraint differs. If $C < XC^a$, then the wired capacity is the bottleneck, and the existence of low rate transmitters does not affect performance, hence there are no performance incentives for handover. If $XC^b < C$, then the wired capacity does not affect the performance results presented in the previous sections. The intermediate case where $XC^a < C < XC^b$ is when the capacity constraint has an impact that depends on other factors. When this holds, we find that case b is beneficial for both access points when the following inequality holds:

$$N1 > \frac{Nx(Nx + N0)rR}{C(Nx(r + R) + 2N0r + \frac{(N0+Nx)ohrR}{pkt}) - (N0 + Nx)rR}. \tag{10}$$

From this equation we see that the number of beneficial scenarios increases, as the wired capacity constraint C increases.

Figure 9 depicts, for 802.11b and $r = 1$ Mbps, the percentage of beneficial scenarios and the corresponding average gain from the perspective of BS1, for different values of the wired capacity constraint C. Experiments for all combinations of $N0, N1, Nx$ are performed, where each of these variables obtains values in $[1, 10]$. As expected, the three zones are clearly visible. When C is less than 1.3 Mbps there are no beneficial scenarios. As C increases above this value, the percentage of beneficial scenarios and the corresponding gain increases until 4 Mbps, when the gain is equal to that of the basic model without wired capacity constraints.

An interesting conclusion is that the wired capacity constraint reduces handover gains when its value is less than the half of the maximum effective throughput of the wireless link. Since the most common wireless protocol used in home

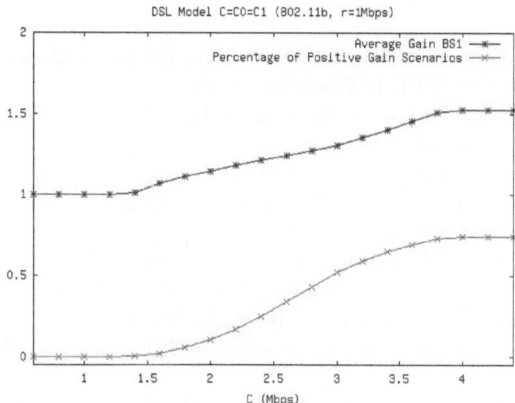

Fig. 9. Effect of the wired capacity constraint on the number of beneficial scenarios and their gain

wireless networks is IEEE 802.11g, which has an effective throughput of about 20-24 Mbps, an ADSL connection which is below 10-12 Mbps would decrease any potential handover gains. In the future, it is likely that we will have a mix of such low rate ADSL connections, together with higher rate connections supported by the deployment of fiber closer to the customers' premises.

7   Conclusion and Future Work

In this paper, we analyzed scenarios where the cooperation between public/home wireless network administrators is motivated solely by the performance improvement for their clients. Using an analytical throughput model with two access points, we computed the gains for both access points, when low rate users are handed-off to the closer access point. Moreover, the analytical model suggests a simple acceptance policy for deciding when handovers are beneficial. The analytical model was compared with simulation results, which verify the accuracy of the model. Finally, we extended the model to include wired capacity constraints, e.g., from ADSL connections.

Ongoing work is investigating the use of priority mechanisms supported by standards such as IEEE 802.11e [14], to control the sharing of the handover gains among the involved access points; such control can also increase the number of beneficial scenarios. We plan to study the details of the implementation of the proposed procedure. We also plan to extend our study with alternative performance metrics, which incorporate notions of fairness in the allocation of throughput among the wireless nodes, and to the case of multiple access points and traffic in the uplink direction. Finally, we plan to investigate the existence of similar performance improvement incentives in the case of wireless mesh networks with overlapping coverage.

References

1. Heusse, M., Rousseau, F., Berger-Sabbatel, G., Duda, A.: Performance anomaly of 802.11b. In: Proc. of IEEE INFOCOM (2003)
2. Akella, A., Judd, G., Seshan, S., Steenkiste, P.: Self-Management in Chaotic Wireless Deployments. WINET Journal 13(6), 737–755 (2007)
3. Siris, V.A., Stamatakis, G.: Optimal CWmin Selection for Achieving Proportional Fairness in Multi-rate 802.11e WLANs: Test-bed Implementation and Evaluation. In: Proc. of ACM WiNTECH (2006)
4. Tan, G., Guttag, J.: Time-based Fairness Improves Performance in Multi-Rate WLANs. In: Proc. of USENIX Annual Technical Conference (2004)
5. Feeney, L.M., Cetin, B., Hollos, D., Kubisch, M., Mengesha, S., Karl, H.: Multi-rate relaying for performance improvement in IEEE 802.11 WLANs. In: Proc. of WWIC (2007)
6. Bahl, V., Chandra, R., Lee, P.P.C., Misra, V., Padhye, J., Rubenstein, D., Yu, Y.: Opportunistic Use of Client Repeaters to Improve Performance of WLANs. In: Proc. of ACM CoNEXT (2008)
7. Kandula, S., Ching-Ju Lin, K., Badirkhanli, T., Katabi, D.: FatVAP: Aggregating AP Backhaul Capacity to Maximize Throughput. In: Proc. of the 5th USENIX Symposium on Networked Systems Design and Implementation (2008)
8. Liu, P., Tao, Z., Narayanan, S., Korakis, T., Panwar, S.: A Cooperative MAC protocol for Wireless LANs. IEEE JSAC 25(2) (2007)
9. Kumar, A., Altman, E., Miorandi, D., Goyal, M.: New Insights from a Fixed Point Analysis of Single Cell IEEE 802.11 WLANs. In: Proc. of IEEE INFOCOM (2005)
10. Kumar, A., Kumar, V.: Optimal Association of Stations and APs in an IEEE 802.11 WLAN. In: Proc. of National Conference on Communications (NCC) (2005)
11. Kasbekar, G., Kuri, J., Nuggehalli, P.: Online Association Policies in IEEE 802.11 WLANs. In: Proc. of 4th Intl. Symposium on Modeling and Optimization in Mobile, Ad Hoc, and Wireless Networks, WiOpt (2006)
12. Kauffmann, B., Baccelli, F., Chaintreau, A., Mhatre, V., Papagiannaki, K., Diot, C.: Measurement-based Self Organization of Interfering 802.11 Wireless Access Networks. In: Proc. of IEEE INFOCOM (2007)
13. Koukoutsidis, I., Siris, V.A.: Access Point Assignment Algorithms in WLANs based on Throughput Objectives. In: Proc. of 6th Intl. Symposium on Modeling and Optimization in Mobile, Ad Hoc, and Wireless Networks (WiOpt) (2008)
14. Mangold, S., Choi, S., Hiertz, G.R., Klein, O.: Analysis of IEEE 802.11e for QoS Support in Wireless LANs. IEEE Wireless Communications 10(6), 40–50 (2003)

Author Index

Anwander, Markus 61
Araujo, Álvaro 73

Barceló, Jaume 1
Bellalta, Boris 1
Braun, Torsten 61
Brzozowski, Marcin 24

Cano, Cristina 1
Cerqueira, Eduardo 121
Conti, Mauro 85

Di Pietro, Roberto 85

Eckhardt, Harald 109

Fafoutis, Xenofon 146
Fallon, Sheila 133

Gabrielli, Andrea 85
Goyeneche, Juan-Mariano de 73
Greifenberg, Janico 97

Harms, Janelle 36

Jacob, Paul 133

Kutscher, Dirk 97

Langendoerfer, Peter 24
Li, Ji 48

Malagón, Pedro 73
Mancini, Luigi Vincenzo 85
Mei, Alessandro 85
Mendes, Paulo 121
Moh, Melody 48
Moh, Teng-Sheng 48
Monteiro, Edmundo 121
Moya, José M. 73
Mueller, Christian M. 109
Murphy, Liam 133

Nürnberger, Stefan 13
Nieto-Taladriz, Octavio 73
Nikolaidis, Ioanis 36
Nolte, Jörg 13

Qiao, Yuansong 133

Sfairopoulou, Anna 1
Sieber, André 13
Sigle, Rolf 109
Siris, Vasilios A. 146

Vallejo, Juan Carlos 73
Veloso, Luis 121

Wagenknecht, Gerald 61
Walther, Karsten 13

Zou, Shoudong 36